U.S. Department of Justice
Office of Justice Programs
National Institute of Justice

SEPT. 03

NIJ

Special REPORT

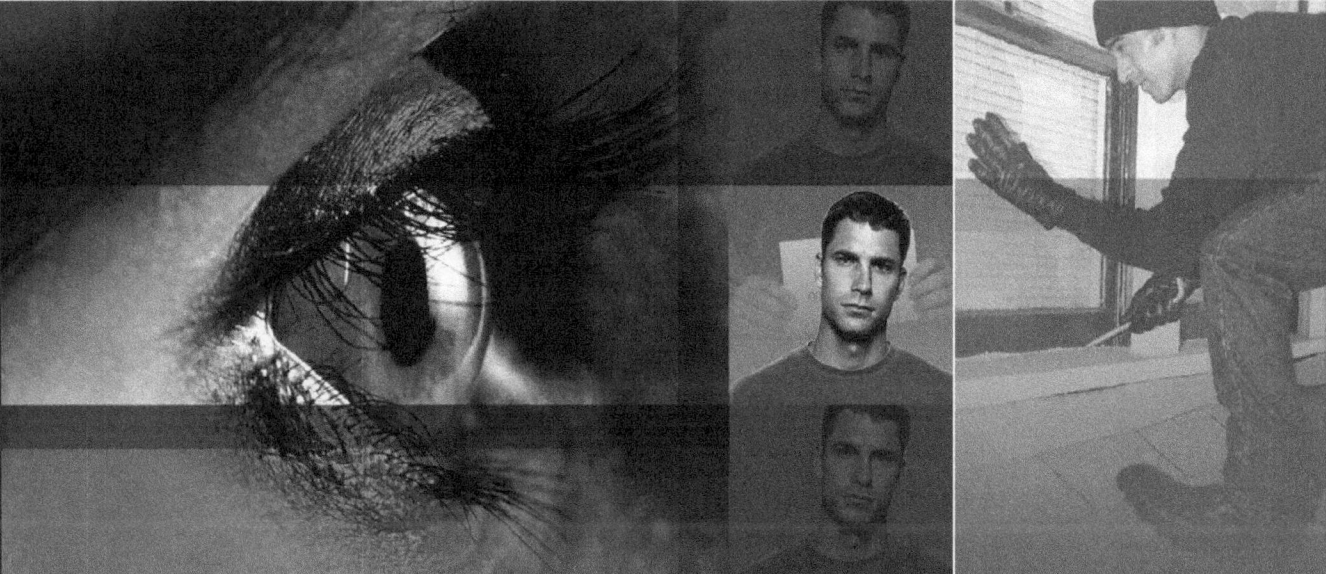

Eyewitness Evidence: A Trainer's Manual for Law Enforcement

U.S. Department of Justice
Office of Justice Programs
810 Seventh Street N.W.
Washington, DC 20531

John Ashcroft
Attorney General

Deborah J. Daniels
Assistant Attorney General

Sarah V. Hart
Director, National Institute of Justice

This and other publications and products of the U.S. Department
of Justice, Office of Justice Programs, National Institute of Justice
can be found on the World Wide Web at the following site:

Office of Justice Programs
National Institute of Justice
http://www.ojp.usdoj.gov/nij

Eyewitness Evidence:

A Trainer's Manual for Law Enforcement

Developed by the

Technical Working Group for Eyewitness Evidence

September 2003

Sarah V. Hart
Director

NCJ 188678

The National Institute of Justice is a component of the Office of Justice Programs, which also includes the Bureau of Justice Assistance, the Bureau of Justice Statistics, the Office of Juvenile Justice and Delinquency Prevention, and the Office for Victims of Crime.

Technical Working Group for Eyewitness Evidence: Training Teams

The Technical Working Group for Eyewitness Evidence (TWGEYEE), a multidisciplinary group of content-area experts from across the United States and Canada, was created by the National Institute of Justice (NIJ) in 1998 to develop recommended procedures for law enforcement use in investigations involving eyewitness evidence. These individuals, led by a Planning Panel composed of distinguished law enforcement, legal, and research professionals, authored the 1999 NIJ Research Report, *Eyewitness Evidence: A Guide for Law Enforcement*.

The TWGEYEE Training Teams were formed in early 2000 as subgroups of TWGEYEE to develop the materials found in this *Trainer's Manual*. Two teams were formed—one to develop a lesson plan and accompanying materials for eyewitness interviewing procedures and another to develop a lesson plan and accompanying materials for identification procedures. Leading law enforcement and research members of the TWGEYEE Planning Panel were designated as team chairs who selected additional team members from the TWGEYEE membership base. These team members, listed below, worked together over a 1-year period to author the *Trainer's Manual*.

Interviewing Team

Chairs

Sgt. Paul Carroll (Ret.)*
Chicago Police Department
Big Pine Key, Florida

Ronald P. Fisher, Ph.D.*
Florida International University
North Miami, Florida

Members

Caterina DiTraglia
Assistant Federal Defender
St. Louis, Missouri

Mark R. Larson*
King County Prosecutor's Office
Seattle, Washington

Roy S. Malpass, Ph.D.
University of Texas at El Paso
El Paso, Texas

Eugene Rifenburg
New York State Police (Ret.)
Oneida Indian Nation Police
Munnsville, New York

Identification Team

Chairs

Det. Lt. Kenneth A. Patenaude*
Northampton Police Department
Northampton, Massachusetts

Gary L. Wells, Ph.D.*
Iowa State University
Ames, Iowa

Members

Cpl. Daniel Alarcon II
Hillsborough County Sheriff's Office
Tampa, Florida

Hon. Michael J. Barrasse
Lackawanna County Judge
Scranton, Pennsylvania

David C. Niblack
Attorney at Law
Washington, D.C.

John Turtle, Ph.D.
Ryerson Polytechnic University
Toronto, Ontario

Alternate

Solomon M. Fulero, Ph.D., J.D.
Sinclair College
Dayton, Ohio

*Planning Panel member

Members of TWGEYEE reviewed the materials contained in this manual and provided valuable input during the preliminary development of the training criteria on which the lesson plans in this manual are built. These members are listed alphabetically below:

First Sgt. Roger Broadbent
Virginia State Police
Fairfax Station, Virginia

Cdr. Ella M. Bully (Ret.)*
Detroit Police Department
Detroit, Michigan

Cpl. J.R. Burton
Hillsborough County Sheriff's Office
Tampa, Florida

Det. Sgt. Chet Bush
Kent County Sheriff's Office
Grand Rapids, Michigan

Carole E. Chaski, Ph.D.*
Institute for Linguistic Evidence
Georgetown, Delaware

Lt. Rosanna Church-Abreo
Texas Department of Public Safety
Special Crime Services
Austin, Texas

Det. Sgt. J. Glenn Diviney (Ret.)
Tarrant County Sheriff's Office
Fort Worth, Texas

James Doyle*
Attorney at Law
Boston, Massachusetts

James Fox
San Mateo County District Attorney
Redwood City, California

Investigations Chief Arlyn Greydanus
Montana Department of Justice
Division of Criminal Investigation
Helena, Montana

Investigator Kathy Griffin
Loveland Police Department
Loveland, Colorado

William Hodgman
Los Angeles County District
  Attorney's Office
Los Angeles, California

Rod C.L. Lindsay, Ph.D.
Queen's University
Kingston, Ontario

Officer Patricia Marshall
Chicago Police Department
Chicago, Illinois

Jeralyn Merritt
Attorney at Law
Denver, Colorado

Melissa Mourges*
New York County District
  Attorney's Office
New York, New York

Patricia Ramirez
Dodge County District Attorney
Juneau, Wisconsin

Det. Edward Rusticus (Ret.)
Kent County Sheriff's Office
Grand Rapids, Michigan

Det. Ray Staley
Kansas City Police Department
Kansas City, Missouri

Lt. Tami Thomas
Atlantic Beach Police Department
Atlantic Beach, North Carolina

Capt. Michael B. Wall
Northampton Police Department
Northampton, Massachusetts

*Planning Panel member

Foreword

Eyewitnesses play a critical role in our criminal justice system. They are often essential to identifying, charging, and ultimately convicting perpetrators of crime and in some cases may provide the sole piece of evidence against those individuals. For these reasons, the value of accurate and reliable eyewitness evidence cannot be overstated.

Cases in which DNA testing has exonerated individuals convicted on the basis of eyewitness testimony tend to make headlines, but in actuality, the frequency of mistaken eyewitness identifications is quite small. The vast majority of eyewitness identifications are accurate and provide trustworthy evidence for the trier of fact.

Recognizing the weight accorded eyewitness evidence by judges and juries, the National Institute of Justice initiated a project in 1998 to research methods to improve the accuracy, reliability, and availability of information obtained from eyewitnesses. The Technical Working Group for Eyewitness Evidence (TWGEYEE), composed of experienced law enforcement investigators, prosecutors, defense lawyers, and psychology researchers, worked together to produce recommendations for the collection and preservation of this vital evidence. These consensus recommendations were included in the 1999 NIJ publication *Eyewitness Evidence: A Guide for Law Enforcement.*

Because of the complex issues associated with identification practices, TWGEYEE recognized that its recommendations may not be feasible in all circumstances. The *Guide*'s recommendations are not legal mandates or policy directives, nor do they represent the *only* correct courses of action. Rather, the recommendations represent a consensus of the diverse views and experiences of the technical working group members, who have provided valuable insight into these important issues. We expect that each jurisdiction will be able to use these recommendations to spark debate and ensure that its practices and procedures are best suited to its unique environment.

Law enforcement personnel can benefit from training based on the procedures recommended in the *Guide*. To assist law enforcement trainers with creating and instructing courses on eyewitness evidence, including the topics of interviewing witnesses and conducting lineups, TWGEYEE has developed the materials included herein. These detailed curriculum plans provide instructors with explanations grounded in research and practical exercises that can enhance learning. For example, this trainer's manual includes a CD–ROM that can be used to guide students through composition of a mock photo lineup. It is our hope that, through these materials, more of our Nation's law enforcement personnel will be trained to work effectively with eyewitnesses and maximize the reliable evidence obtained from them, to the benefit of criminal case prosecutions.

NIJ extends its appreciation to the Baltimore County (Maryland) Police Department and the Mid-Atlantic Regional Community Policing Institute (MARCPI) for their willingness to test the training materials included in this manual—as well as their time and effort in doing so—in the interest of delivering a better product to the Nation's law enforcement community. NIJ is particularly grateful for the support of those who orchestrated and carried out the pilot testing of this manual: Captain Howard Hall, Sergeants Theresa McQuaid, Samuel Hannigan, and Melvin Teal, and Officers Scott

Leonard and James Moss of the Baltimore County Police Department Training Academy; Director William Rogers of MARCPI; and Karl Bickel of the U.S. Department of Justice's Office of Community Oriented Policing Services. Thanks also are extended to the many law enforcement and legal practitioners from the Baltimore-Washington area who attended the pilot training course and provided valuable suggestions to refine these materials. Special thanks are extended to Detective Lieutenant Kenneth Patenaude of the Northampton (Massachusetts) Police Department for his selfless efforts in developing the CD–ROM that accompanies this manual. Finally, we thank the members of the TWGEYEE Training Panel for their extensive efforts on this project and their dedication to strengthening the value of eyewitness evidence in the criminal justice system, as well as the original TWGEYEE members, who continued their commitment to this project by reviewing and commenting on this manual throughout its development stages. We believe that the overall improvement in professional practices that will result from this project will ultimately lead to stronger evidence for criminal cases and reliable verdicts.

The TWGEYEE members have dedicated their work to the memory of David C. Niblack—steadfast working group member, devoted advocate, and also, friend.

Sarah V. Hart
Director, National Institute of Justice

Contents

Trainers should carefully read this
manual in its entirety before attempting
to teach this material.

About This Manual

Purpose of the Trainer's Manual

This manual is written for law enforcement trainers to accompany *Eyewitness Evidence: A Guide for Law Enforcement* (hereafter, the *Guide*). It is presumed that the law enforcement students in the course will each have a copy of (or access to) the *Guide*. This manual and the accompanying CD–ROM are designed to facilitate trainers' teaching of the *Guide* in two ways.

First, this manual provides much of the context for understanding *why* the procedures described in the *Guide* for the collection and preservation of eyewitness evidence may enhance the reliability of this evidence. Although the procedures described in the *Guide* are relatively easy to follow, it is useful for trainers to understand why procedures are important and to communicate these reasons to the students.

Second, the CD–ROM includes a multimedia presentation that can supplement trainers' discussion. The presentation includes a number of exercises and demonstrations to promote students' understanding of the material and to make the training sessions more interactive and interesting for the students.

Development of the Trainer's Manual

This manual was developed by a panel of law enforcement practitioners, psychology researchers, prosecutors, and defense lawyers who served as members of the Technical Working Group for Eyewitness Evidence (TWGEYEE). The panel used a consensus-building process similar to that employed in developing the *Guide* to write the instructor's notes and explanations included in this manual. All material was reviewed by TWGEYEE as well as a network of national organizations, law enforcement agencies, and individuals concerned with the training of investigators. Comments from these reviewers were considered by the panel and incorporated as appropriate.

The sample lesson plans and materials included in this manual were pilot tested by police training instructors in an authentic classroom environment. This pilot testing was completed in September 2000 at the Baltimore County (Maryland) Police Department Training Academy as part of a 2-day training course for law enforcement and legal practitioners and trainers. Instructor and student evaluation forms were completed, and feedback sessions with the authors of the manual were held at the end of each day to assess the utility and effectiveness of the lesson plans and training materials. Further refinements to this manual were accomplished as a result of the pilot testing.

Organization of the Trainer's Manual

The manual is divided into two sections: Interviewing and Identification. These sections provide sample lesson plans that relate directly to the interview and identification procedures contained in the *Guide*. In these sections, the procedural portion of the *Guide* is reprinted in black, with explanations for the instructor's reference included in blue following the procedures themselves. Instructors should describe the rationale for each procedure at the time each procedure is discussed. Information in the "notes" column (also printed in blue) suggests exercises, demonstrations, and discussion ideas that correspond to the procedures, as well as key points that should be highlighted during classroom discussion. The notes column also includes references to the contents of the accompanying CD–ROM.

The CD–ROM includes a multimedia presentation that follows the procedural content of the *Guide*. Its use is highly recommended because it includes audio/visual aides and exercises that can be effective teaching tools. Instructors are, of course, encouraged to incorporate their own materials as desired.

The CD–ROM also includes an Adobe Acrobat (.pdf) file of the *Guide*, which can be printed and photocopied for students in the event that actual *Guides* are not available. An executable version of Adobe Acrobat Reader (.exe) is included on the CD–ROM for instructors' convenience.

Use of the Trainer's Manual

The material in this manual is designed either to supplement existing training programs or to be used on its own. Each component of the Interviewing and Identification sections may be taught independently as needed for different audiences (e.g., dispatchers, first responders, investigators, interviewers, lineup administrators).

When conducting a training course using this manual, the instructors' explanations in blue text should be used to clarify for students the reasoning behind each procedure. The instructors' explanations often include practical examples of what can go wrong if procedures are not followed, which help to "drive home" the importance of each step for students. Key points and clarifications are highlighted in the notes column to draw instructors' attention to critical information. Instructors should use this information to facilitate students' learning, understanding, and skills. The notes column, as the name implies, is also useful for instructors to make notes to themselves for future training sessions.

Instructors should keep in mind that students will have access to the *Guide* itself, rather than this manual. As a practical consideration (particularly in courses operating within strict time limitations), it is more important to emphasize procedural *reasoning and examples* as included in the instructors' explanations than to read verbatim the text of the *Guide*. In particular, "Principle," "Policy," and "Summary" statements need not be read to the class. A simple summary of each section's importance can be provided by the instructor.

Instructors may note that some departures have been made from the text of the *Guide* in certain areas of this manual. These departures reflect that the performance of every procedure may not be feasible or appropriate in all jurisdictions or in all situations. Instructors may choose to mention these caveats to students as appropriate.

The CD–ROM presentation that accompanies this manual can be used with a laptop computer and LCD (liquid crystal display) projector to enhance training. The audio/visual aides and exercises included in the presentation are referenced in the manual's notes column at points where their use is most beneficial. Links to audio files appear as numbered loudspeaker icons on the presentation slides. Video files appear as separate slides that contain the first frame of the video. To play the audio and video files, the instructor need only move the cursor over the icon or frame.

Instructors should familiarize themselves thoroughly with this manual before attempting to train law enforcement students on the *Guide*'s procedures. Instructors are encouraged to direct questions and suggestions regarding both the *Guide* and this manual to the National Institute of Justice (*askost@ojp.usdoj.gov*). Such feedback from law enforcement trainers will be valuable for any future updates of these materials.

The procedures contained in this manual are intended as examples of suggested practices for the collection and preservation of eyewitness evidence.

Departmental, logistical, or budget/staff limitations or legal conditions may make the use of particular procedures contained herein impracticable.

Instructors are encouraged to modify these materials as appropriate for their jurisdiction and classroom presentation.

Sample Lesson Plan: Interviewing

Role-playing is used in this lesson plan as a tool to facilitate students' learning of effective interview techniques. With classes of up to 25 students, use role-playing exercises in which all students rotate through the roles of interviewer/call-taker and witness. With classes of more than 25 students, select two volunteers to go through role-playing exercises for the class to observe.

Instruct the student playing the role of the witness to think of an actual event that he/she experienced approximately 6 months earlier. Using memory of an actual event in these exercises will more effectively demonstrate the utility of the interview procedures contained herein.

To maximize the educational benefit of the exercise, the procedural point being demonstrated should be explained prior to the role-playing exercise.

Show Slide 1

Section I. Initial Report of the Crime/First Responder (Preliminary Investigator)

Show Slide 2

A. Answering the 9—1—1/Emergency Call (Call-Taker/Dispatcher)

Principle: As the initial point of contact for the witness, the 9–1–1/emergency call-taker or dispatcher should obtain and disseminate, in a non-suggestive manner, complete and accurate information from the caller. This information can include the description/identity of the perpetrator of a crime. The actions of the call-taker/dispatcher can affect the safety of those involved as well as the entire investigation.

NOTE:
Instructors are encouraged to play an audiotaped example of an actual 9–1–1 call for classroom discussion.

Policy: The call-taker/dispatcher should answer each call in a manner conducive to obtaining and disseminating accurate information regarding the crime and the description/identity of the perpetrator.

Procedure: During a 9–1–1/emergency call—after obtaining preliminary information and dispatching police—the call-taker/dispatcher should—

Show Slide 3

1. Assure the caller the police are on the way.

   👁 This will help to calm the caller so he/she can focus on providing information.

Show Slides 4–6:
Give examples of open-ended and closed-ended questions.

IMPORTANT:
Emphasize that questioning should be primarily open-ended; use closed-ended questions *only as necessary*.

2. Ask open-ended questions (e.g., "What can you tell me about the car?") and augment with closed-ended questions (e.g., "What color was the car?").

👁 An open-ended question allows for an unlimited response from the witness in his/her own words (e.g., "What can you tell me about the perpetrator?" or "Tell me in your own words what happened.") Open-ended questions allow the caller to play an active role, thereby generating a greater amount of unsolicited information. Open-ended responses also tend to be more accurate and promote more effective listening on the part of the call-taker. The call-taker also is less likely to lead the witness when framing questions in this manner.

👁 A closed-ended question, in contrast, limits the amount or scope of information that the witness can provide (e.g., "Did the perpetrator have a beard?" or "What color was the car?"). Although it is preferable to use open-ended questioning, the call-taker should follow with more directed questions if the caller is unresponsive to open-ended questions or provides imprecise responses. If, for example, when answering an open-ended question, the witness states that the perpetrator had a weapon, the call-taker should ask the witness what type of weapon it was.

Show Slide 7:
Give examples of leading questions.

3. Avoid asking suggestive or leading questions (e.g., "Was the car red?").

👁 Leading questions suggest an answer and may distort the caller's perception or memory. The call-taker needs to determine only what the caller knows, uninfluenced by what the call-taker might expect or know from other sources. For example, the call-taker may have been informed by another caller that the car was red, but should not ask, "Was it a red Honda?" Or, if the call-taker receives a call about a domestic situation, the call-taker should not ask, "Did your husband hit you?" but should ask, "What happened?" or "What is going on now?"

Show Slide 8

4. Ask if anything else should be known about the incident.

👁 This gives the caller a chance to recall and report any extra information they may have and also contributes to the safety of responding officers.

5. Transmit information to responding officer(s).

👁 This is necessary for officer safety. Complete information in the hands of responding officers also can result in faster resolution of the incident.

6. Update officer(s) as more information comes in.

Summary: The information obtained from the witness is critical to the safety of those involved and may be important to the investigation. The manner in which facts are elicited from a caller can influence the accuracy of the information obtained.

B. Investigating the Scene [Preliminary Investigating Officer]

Principle: Preservation and documentation of the scene, including information from witnesses and physical evidence, are necessary for a thorough preliminary investigation. The methods used by the preliminary investigating officer have a direct impact on the amount and accuracy of the information obtained throughout the investigation.

Policy: The preliminary investigating officer should obtain, preserve, and use the maximum amount of accurate information from the scene.

Procedure: After securing the scene and attending to any victims and injured persons, the preliminary investigating officer should—

1. Identify the perpetrator(s).

   👁 Question persons present at the scene to obtain a description of the perpetrator if still at large.

   a. Determine the location of the perpetrator(s).

      👁 Determine location of perpetrator if it is known or direction/means of travel if the perpetrator fled the scene.

   b. Detain or arrest the perpetrator(s) if still present at the scene.

2. Determine/classify what crime or incident has occurred.

3. Broadcast an updated description of the incident, perpetrator(s), and/or vehicle(s).

   👁 New information can affect resource deployment and type of response (e.g., personnel, support services, or equipment needed).

4. Verify the identity of the witness(es).

   👁 Witnesses will need to be contacted later. Obtain and document valid forms of identification and contact information for each witness. List all witnesses in a written report.

Show Slide 9

Conduct role-playing exercises. Have students rotate through both roles (call-taker and caller).

Show Slide 10

Show Slide 11

Show Slide 12

IMPORTANT:
Be sure to clarify
the reasoning behind
this procedure; give
example(s).

5. Separate witnesses and instruct them to avoid discussing details of the incident with other witnesses.

👁 Witnesses should not hear others' accounts because they may be influenced by that information. Independent witness statements can corroborate other witnesses' statements and other evidence in the investigation. The following example demonstrates how failure to separate witnesses could mislead an investigation: Suppose that a crime is committed by a perpetrator who is clean shaven. If one witness incorrectly states that the perpetrator had a beard and other witnesses overhear that statement, it could lead them to report that they also saw a beard when in fact they did not. This would direct investigators to search for a bearded suspect.

👁 It also may be helpful to ascertain whether witnesses have spoken with each other about the incident prior to being separated.

6. Canvass area for other witnesses.

👁 Witnesses may be reluctant to come forward for any number of reasons or may have departed the scene before law enforcement personnel arrived. Also, other persons in the vicinity, such as neighbors or shopkeepers, may have heard or seen something that could assist in the investigation.

Show Slide 13

Summary: The preliminary investigation at the scene forms a sound basis for the accurate collection of information and evidence during the followup investigation.

Show Slide 14

C. Obtaining Information From Witness(es)

Principle: The manner in which the preliminary investigating officer obtains information from a witness impacts the amount and accuracy of that information.

Policy: The preliminary investigating officer should obtain and accurately document and preserve information from the witness(es).

Procedure: When interviewing a witness, the preliminary investigating officer should—

Play Audio Cut 1
(example of poor
rapport development)
and Audio Cut 2
(example of good
rapport development).

1. Establish rapport with the witness.

👁 The development of rapport between the witness and investigator will make the witness more comfortable during the interview process. Comfortable witnesses will generally provide more information. In the course of developing rapport with the witness, the investigator can learn about the witness's communication style (e.g., how the witness describes everyday events compared with how the witness describes the incident).

2. Inquire about the witness's condition.

👁 A simple question, such as "How are you doing?" will not only contribute to rapport development, but it can alert the investigator to physical or mental conditions (e.g., intoxication, medication, shock) that could potentially impair the witness's ability to recall or report information effectively.

3. Use open-ended questions (e.g., "What can you tell me about the car?") and augment with closed-ended questions (e.g., "What color was the car?"). Avoid leading questions (e.g., "Was the car red?").

Show Slide 15

👁 An open-ended question allows for an unlimited response from the witness in his/her own words (e.g., "What can you tell me about the perpetrator?" or "Tell me in your own words what happened"). Open-ended questions allow the witness to play an active role, thereby generating a greater amount of unsolicited information. Open-ended responses also tend to be more accurate and promote more effective listening on the part of the investigator. The investigator also is less likely to lead the witness when framing questions in this manner.

Show Slide 16
EXERCISE:
Have students convert closed-ended questions to open-ended questions.

Acceptable student responses include—

1. *"What did his hair look like?"* (allows for answers about style, length, color, texture).

👁 A closed-ended question, in contrast, limits the amount or scope of information that the witness can provide (e.g., "Did the perpetrator have a beard?" or "What color was the car?"). Although it is preferable to use open-ended questioning, the investigator should follow with more directed questions if the witness is unresponsive to open-ended questions or provides imprecise responses. If, for example, when answering an open-ended question, the witness states that the perpetrator was dressed in "shabby" clothing, the investigator should ask the witness to elaborate on the type of clothing (e.g., "What do you mean by 'shabby'?").

2. *"What was he wearing?"* (allows for answers about the perpetrator's clothing as a whole, including pants, shoes, hat, shirt, jacket, jewelry, etc., and allows for details such as "ragged" or "shiny").

👁 For each new topic of information being sought, the investigator should begin with open-ended questions and augment them with closed-ended questions if necessary. For example, if, after having elicited all information from the witness about the perpetrator, the next topic of information is the getaway car, the investigator should begin this line of inquiry with open-ended questions about the car.

3. *"What did his face look like?"* (allows for answers about facial features and other details such as scars or unusual aspects of the perpetrator's face).

👁 Leading questions suggest an answer and may distort the witness's perception or memory. The investigator needs to determine only what the witness knows, uninfluenced by what the investigator might expect or know from other sources. For example, the investigator may have been informed by another witness that the car was red, but should not ask, "Was it a red Honda?"

Show Slide 17

4. Clarify the information received with the witness.

   👁 Asking the witness about what they have reported ensures that the information has been understood and accurately recorded.

5. Document information obtained from the witness, including the witness's identity, in a written report.

   👁 This information will be necessary when the witness is contacted for a followup interview.

6. Encourage the witness to contact investigators with any further information.

   👁 Witnesses will often remember additional, useful information after an interview. Remind the witness that any information, no matter how trivial it may seem, is important. For example, if the witness later remembers that the perpetrator drank from a soft drink can at the scene, there could be fingerprints or saliva on the can. Additionally, in such cases as sexual assault or arson, the witness may later recall or recognize a distinct smell that was either on the perpetrator (such as cologne) or at the scene (such as gasoline) that could be useful in developing leads.

Show Slide 18

7. Encourage the witness to avoid contact with the media or exposure to media accounts concerning the incident.

   👁 Media information may contaminate the witness's memory. Media requests for a story or offers of compensation may encourage a witness to fabricate information.

IMPORTANT:
Be sure to clarify the reasoning behind this procedure; give example(s).

8. Instruct the witness to avoid discussing details of the incident with other potential witnesses.

   👁 Witnesses should not hear others' accounts because they may be influenced by that information. The independence of witnesses is important to see if the information they have provided is consistent with other witnesses' statements and other evidence in the investigation. As an example of the importance of independent sources for corroboration, suppose you wanted to corroborate a crime report that appeared in a local newspaper. Finding a second copy of that newspaper that reports the same story does not corroborate the first newspaper story because they are from the same source. Proper corroboration requires that the source be a second, independent news report.

Show Slide 19

Conduct role-playing exercises for obtaining information from witnesses.

Summary: Information obtained from the witness can corroborate other evidence (e.g., physical evidence, accounts provided by other witnesses) in the investigation. Therefore, it is important that this information be accurately documented in writing.

Section III. Procedures for Interviewing the Witness by the Followup Investigator

Show Slide 20

A. Preinterview Preparations and Decisions

Show Slide 21

Principle: Preparing for an interview maximizes the effectiveness of witness participation and interviewer efficiency.

Policy: The investigator should review all available witness and case information and arrange an efficient and effective interview.

Procedure: Prior to conducting the interview, the investigator should—

1. Review available information.

   👁 This information may include police reports and crime scene information. It is important for the interviewer to have all information relevant to the case prior to conducting the interview so that the interview can be tailored to elicit the maximum amount of information from the witness.

2. Plan to conduct the interview as soon as the witness is physically and emotionally capable.

   👁 Once the witness is capable, any delay in conducting the interview should be minimized as there will be less detailed information available as time goes on.

3. Select an environment that minimizes distractions while maintaining the comfort level of the witness.

   Show Slide 22

   👁 Distractions will interrupt the witness's memory retrieval. Avoid interviewing the witness in an environment where distractions are more likely to occur, such as a place of business. This should be determined with the witness to accommodate his/her schedule and needs.

4. Ensure resources are available (e.g., notepad, tape recorder, camcorder, interview room).

   👁 Secure these items prior to the interview so the interview will not be interrupted.

5. Separate the witnesses.

   Show Slide 23

   👁 Independent witness statements can be used as corroboration/ confirmation. Witnesses should not hear others' statements because they may be influenced by that information.

Clarify that this procedure involves general law enforcement contact, *not* contact related to this case. The purpose of this procedure is to assess the witness's credibility.

Show Slide 24

Show Slide 25

Show Slide 26

IMPORTANT:
Clarify that this procedure involves contact related to *witnessing the incident*. Do *not* ask the witness about his/her criminal record (this type of information should have been obtained during preparation for the interview).

6. Determine the nature of the witness's prior law enforcement contact.

👁 Prior law enforcement contact may include an arrest record, prior victimization, warrants, or any relationship to/with law enforcement personnel. This information can help put any information obtained from the witness into context for the purpose of assessing witness credibility and/or reliability. It also can assist later in rapport development.

Summary: Performing the above preinterview preparations will enable the investigator to elicit a greater amount of accurate information during the interview, which may be critical to the investigation.

B. Initial (Preinterview) Contact With the Witness

Principle: A comfortable witness provides more information.

Policy: Investigators should conduct themselves in a manner conducive to eliciting the most information from the witness.

Procedure: On meeting with the witness but prior to beginning the interview, the investigator should—

1. Develop rapport with the witness.

👁 The development of rapport between the witness and interviewer will make the witness more comfortable during the interview process. Comfortable witnesses will generally provide more information. In the course of developing rapport with the witness, the interviewer can learn about the witness's communication style (e.g., how the witness describes everyday events as compared with how the witness describes the incident). For example, if the witness appears nervous during the rapport development phase, the interviewer should not necessarily interpret nervous responses to later questions as being fabrications.

2. Inquire about the nature of the witness's prior law enforcement contact related to the incident.

👁 Prior law enforcement contact related to the incident includes interviews by other officers at the scene, participation in a show-up and with whom, and so forth. This information can help put the witness's comments into context. Do not ask about prior criminal record at this time. The interviewer should ask the witness if he/she has heard any other accounts of the incident (e.g., through the media, from other witnesses).

3. Volunteer no specific information about the suspect or case.

👁 Telling witnesses facts about the suspect or case may influence their memories of the incident. The interviewer must ensure that information from the witness is based only on the witness's memory and not on any information gleaned from the interviewer.

Summary: Establishing a cooperative relationship with the witness likely will result in an interview that yields a greater amount of accurate information.

Show Slide 27

C. Conducting the Interview

The following is a summary of the order in which interviewing concepts should be instructed for maximum benefit. These concepts are more thoroughly discussed in *Memory Enhancing Techniques for Investigative Interviewing* (Fisher and Geiselman, 1992) (see Further Reading). After these concepts are explained, the 12 most important procedural points are listed as they appear in the *Guide*.

There are four basic principles of interviewing cooperative witnesses:

👁 Social dynamics between the interviewer and witness.

👁 Facilitation of the witness's memory and thinking.

👁 Communication between the interviewer and witness.

👁 Sequence of the interview.

Social Dynamics Between the Interviewer and Witness

Two goals are critical to establishing appropriate social dynamics:

👁 Maintain or reestablish rapport with the witness.

👁 Encourage the witness to actively and voluntarily report information, rather than passively respond to the interviewer's questions.

Establishing rapport

When seeking to obtain information of a personal or intimate nature from a witness, establishing a personal relationship with the witness gains his/her trust. Rapport development will help the witness to feel more comfortable conveying personal information. It can be accomplished by personalizing the interview and by developing and communicating empathy.

Show Slide 28:

Play Audio Cuts 3 and 4 (examples of two contrasting interview techniques): Ask students to hypothesize as to why one set of techniques works better than the other.

IMPORTANT:
Explain the four basic principles of interviewing and *why they are essential*. Provide examples of how the associated procedures can impact the information obtained.

Show Slide 29:
IMPORTANT:
The following information on the four principles should be conveyed or read to the class. Include examples that are supported by audio cuts.

- *Show understanding and concern.* This can be accomplished by asking about the witness's health, empathizing with the witness's situation, avoiding judgmental comments, and establishing common ground with the witness.

- *Personalize the interview.* The interviewer should treat the witness as an individual and not as a mere statistic. This can be accomplished by avoiding pre-memorized questions that sound programmed or artificial (e.g., "Is there anything you can tell me that would further assist this investigation?") and referring to the witness by his/her name.

- *Listen actively.* The interviewer should ask interactive questions that follow up on the witness's previous responses, repeat witness's concerns, lean forward, and make eye contact.

Active generation of information

Play Audio Cuts 5*, 6**, and 7*** (examples of poor technique)

The witness should be encouraged to volunteer information without prompting.* Because the witness, rather than the interviewer, possesses the relevant information, the witness should be mentally active during the interview and generate information, as opposed to being passive and waiting until the interviewer asks the appropriate question before answering. The interviewer can encourage the witness to be mentally active by directly requesting this activity or by asking open-ended questions. An open-ended question allows for an unlimited, narrative response from the witness (e.g., "What can you tell me about the perpetrator?").** The interviewer should avoid interrupting the witness's answer to an open-ended question.***

Encouraging the witness to actively generate information can be accomplished by—

- *Stating expectations.* This is important because witnesses may not know what to expect or may have incorrect expectations of their role in the interview. The interviewer should state explicitly that the witness is expected to volunteer information.

- *Asking open-ended questions.* These questions allow the witness to do most of the talking during the interview and can make the witness feel more in control.

- *Avoiding interruptions.* Interrupting the witness during his/her answer discourages the witness from playing an active role and disrupts his/her memory. Rather than interrupt, the interviewer should make a note and follow up at a later time with any questions that arise during a witness's narration.

👁 *Allowing pauses.* It is important to allow for pauses after the witness stops speaking and before continuing to the next question. These periods of silence allow the witness to collect his/her thoughts and continue responding, thereby providing a greater amount of information.

Facilitation of the Witness's Memory and Thinking

Much of the information about the incident is stored in the witness's mind. For the witness to remember these events, he/she must concentrate and search through memory efficiently. The interviewer can promote information retrieval in several ways:

👁 *Minimize distractions.* The interviewer should ensure that physical distractions, such as noise or the presence of other persons, are minimized. In addition, the interviewer can encourage the witness to block out these distractions by closing his/her eyes and concentrating on the memory.

👁 *Encourage the witness to mentally recreate the incident.* The interviewer can promote the witness's efficient recollection of the incident by instructing the witness to mentally recreate the circumstances surrounding the incident (e.g., think about his/her thoughts or feelings at the time of the incident).

👁 *Tailor questions to the witness's narrative.* Because the witness is the source of information, the interviewer's questions should be tailored to the witness's current thoughts and narrative. For example, if the witness is thinking or talking about the perpetrator's face, the questions should be about the face and not about other aspects of the incident, such as a license plate.* The interviewer should try to understand what aspect of the incident the witness is thinking about. Based on this inference, the interviewer should ask an open-ended question about that topic and then follow up with nonleading, closed-ended questions related to that topic. A closed-ended question is specific and limits the witness's response to one or two words (e.g., "How tall was he?"). When asking closed-ended questions, the interviewer must ensure that the questions are nonleading. A leading question suggests an answer to the witness (e.g., "Was his hair blond?").

Communication Between the Interviewer and Witness

The interviewer has investigative needs to solve the crime and the witness possesses relevant knowledge about the details of the crime. Both individuals need to communicate to each other this information. Otherwise, information may not be fully or effectively reported.

Play Audio Cut 8 (example of good technique)

Conduct role-playing exercises focusing on social dynamics and get feedback.

Show Slide 30

* Play Audio Cut 9 (example of poor technique)

Conduct role-playing exercises focusing on facilitation of the witness's memory and thinking and get feedback.

Show Slide 31

Conduct role-playing exercises focusing on communication and get feedback.

Show Slide 32

The interviewer should convey investigative needs (i.e., the types of information he/she is looking for) to the witness. The investigator needs the witness to report the event in more detail than would be conveyed in normal conversation. The investigator should explain this need for detail to the witness to ensure the witness is fully aware of how to provide the description.

Witnesses may have a very good memory of the incident but fail to communicate the knowledge effectively. Therefore, the interviewer should try to facilitate the witness's conversion of memory into effective communication. This can be accomplished by encouraging nonverbal responses (e.g., drawings, gestures) to supplement verbal descriptions as appropriate. The interviewer should also encourage the witness to report all information and not edit his/her thoughts. However, the witness should be cautioned not to guess simply to please the interviewer. It is preferable that the witness state, "I don't know," or indicate that he/she is uncertain about a given answer.

Sequence of the Interview

To be effective in obtaining the maximum amount of information from a witness, the interview should be conducted in stages. The structure of the interview is first designed to calm the witness and gain his/her trust. The interview should continue with general instructions provided by the interviewer, followed by the witness's narrative, and then relevant, probing questions by the interviewer. (Note: Ideally, information should be gathered using primarily open-ended questions. More specific, closed-ended questions should be used only when the witness fails to provide a clear or complete response.) The interview is then closed, leaving lines of communication open between the interviewer and witness.

The following is an example of a sequence to conduct the interview:

1. Attempt to minimize the witness's anxiety.

2. Establish and maintain rapport.

3. Encourage the witness to take an active role in the interview.

4. Request a "free narrative" description of the incident.

5. Ask the witness to mentally recreate the circumstances of the incident.

6. Ask followup questions to elicit additional information related to the witness's narration.

7. Review your notes and other materials.

8. Ask the witness, "Is there anything else I should have asked you?"

9. Close the interview.

To review, the course structure should be based on the concepts described above and follow the outline: Social Dynamics, Memory/Thinking, Communication, and Sequence. At the end of each of the four sections, role-playing exercises should be conducted. Following are the key interviewing procedures as they appear in the *Guide*.

Principle: Interview techniques can facilitate witness memory and encourage communication both during and following the interview.

Policy: The investigator should conduct a complete, efficient, and effective interview of the witness and encourage postinterview communication.

Procedure: During the interview, the investigator should—

1. Encourage the witness to volunteer information without prompting.

   👁 This allows the witness to maintain an active role in the interview. Unprompted responses tend to be more accurate than those given in response to an interviewer's questioning. Use a structured format (e.g., fill-in-the-blank form) only after you have collected as much information as possible from open-ended questions.

2. Encourage the witness to report all details, even if they seem trivial.

   👁 Sometimes the witness may withhold relevant information because he/she thinks it is unimportant or out of order. All information the witness provides is important.

3. Ask open-ended questions (e.g., "What can you tell me about the car?") and augment with closed-ended, specific questions (e.g., "What color was the car?").

   👁 Open-ended questions allow the witness to play an active role, thereby generating a greater amount of unsolicited information. Open-ended responses also tend to be more accurate and promote more effective listening on the part of the interviewer. The interviewer also is less likely to lead the witness when framing questions in this manner. Ideally, information should be gathered using primarily open-ended questions. More specific, closed-ended questions should be used only when the witness fails to provide a clear or complete response.

4. Avoid leading questions (e.g., "Was the car red?").

   👁 Leading questions suggest an answer and may distort the witness's memory.

Conduct role-playing exercises or practice interviews and get feedback. Use civilians as witnesses when possible.

Show Slide 33

Show Slide 34: Reiterate the importance of using primarily open-ended questions.

Show Slide 35	5. Caution the witness not to guess. ◉ Witnesses, particularly child witnesses, may guess in an attempt to please the interviewer. Instruct the witness to state any uncertainty he/she may feel concerning an answer. 6. Ask the witness to mentally recreate the circumstances of the event (e.g., "Think about your feelings at the time"). ◉ Recreating the circumstances of the event makes memory more accessible. Instruct the witness to think about his/her thoughts and feelings at the time of the incident. 7. Encourage nonverbal communication (e.g., drawings, gestures, objects). ◉ Some information can be difficult to express verbally. Witnesses, especially children and witnesses responding in other than their first language, may have difficulty with verbal expression. Witnesses' recall can be enhanced by encouraging them to draw diagrams of the crime scene, perpetrator's scars, and so forth or to use gestures to demonstrate actions.
Show Slide 36 **IMPORTANT:** Emphasize the usefulness of allowing "pauses."	8. Avoid interrupting the witness. ◉ Interrupting the witness during an answer discourages the witness from playing an active role and disrupts his/her memory. Do not immediately continue questioning when a witness pauses after an answer. During a pause, the witness may be collecting his/her thoughts and could continue to provide information, if provided ample time. 9. Encourage the witness to contact investigators when additional information is recalled. ◉ Witnesses will often remember additional, useful information after the interview. Remind the witness that any information, no matter how trivial it may seem, is important.
Show Slide 37	10. Instruct the witness to avoid discussing details of the incident with other potential witnesses. ◉ Witnesses should not hear others' accounts because they may be influenced by that information. The independence of witnesses is important for corroboration of the information they have provided with other witnesses' statements and other evidence in the investigation.

11. Encourage the witness to avoid contact with the media or exposure to media accounts concerning the incident.

👁 Media information may contaminate the witness's memory. Media requests for a story or offers of compensation may encourage witnesses to fabricate information.

12. Thank the witness for his/her cooperation.

👁 This reinforces the rapport that has been developed and the interviewer's commitment to the witness, encouraging the witness to continue to cooperate.

Summary: Information elicited from the witness during the interview may provide investigative leads and other essential facts. The above interview procedures can enable the witness to provide an accurate, complete description of the event and encourage the witness to report later recollections. Witnesses commonly recall additional information after the interview that may be critical to the investigation.

Show Slide 38

D. Recording Witness Recollections

Show Slide 39

NOTE:
These procedures are conducted *with the witness.*

Principle: The record of the witness's statements accurately and completely reflects all information obtained and preserves the integrity of this evidence.

Policy: The investigator should provide complete and accurate documentation of all information obtained from the witness.

Procedure: During or as soon as reasonably possible after the interview, the investigator should—

1. Document the witness's statements (e.g., audio or video recording, stenographer's documentation, witness's written statement, written summary using witness's own words).

👁 Documentation is imperative in the instance that the witness cannot be located later. Use of the witness's own words ensures that the information is recorded accurately. Additionally, in some jurisdictions, the witness's statement must be signed to be admissible in court.

2. Review written documentation; ask the witness if there is anything he/she wishes to change, add, or emphasize.

Show Slide 40

👁 This is useful for clarifying the information received from the witness to ensure the information has been recorded accurately. This also provides an extra opportunity for witnesses to remember additional information.

Show Slide 41

Summary: Complete and accurate documentation of the witness's statement supports a successful investigation and any subsequent court proceedings.

Show Slide 42

E. Assessing the Accuracy of Individual Elements of a Witness's Statement

NOTE:
These procedures are conducted after the interview, *without the witness.*

Principle: Point-by-point consideration of a statement may enable judgment on which components of the statement are most accurate. Each piece of information recalled by the witness may be remembered independently of other elements.

Policy: The investigator should review the individual elements of the witness's statement to determine the accuracy of each point.

Procedure: After conducting the interview, the investigator should—

1. Consider each individual component of the witness's statement separately.

   👁 A witness may not have information about all elements of an incident. Thus, some recollections may be correct while others may be incorrect.

Show Slide 43:
Step 2 examines the *internal consistency* of the statement.

2. Review each element of the witness's statement in the context of the entire statement. Look for inconsistencies within the statement.

   👁 Note any inconsistencies for future reference. Also, note that the inconsistency of one element with another does not imply that the entire statement is inaccurate.

Step 3 examines the *external consistency* of the statement as it relates to other information obtained in the case investigation.

3. Review each element of the statement in the context of evidence known to the investigator from other sources (e.g., other witnesses' statements, physical evidence).

   👁 Note any inconsistencies between the witness's statement and other information. These inconsistencies can be useful in assessing the accuracy of elements of witness statements as well as in directing the investigation.

Show Slide 44

Summary: Point-by-point consideration of the accuracy of each element of a witness's statement can assist in focusing the investigation. This technique avoids the common misconception that the accuracy of an individual element of a witness's description predicts the accuracy of another element.

F. Maintaining Contact With the Witness

Show Slide 45

Principle: The witness may remember and provide additional information after the interview has concluded.

Policy: The investigator should maintain open communication to allow the witness to provide additional information.

Procedure: During postinterview, followup contact with the witness, the investigator should—

1. Reestablish rapport with the witness.

   👁 The investigator should ask the witness about something personal that follows up on his/her previous contact with the witness (e.g., "Has your arm healed?"). Witnesses will continue to provide information to investigators with whom they have a continuous positive relationship.

2. Ask the witness if he/she has recalled any additional information.

   👁 This reinforces the idea that the witness is an active part of the investigation. Witnesses generally recall additional information following the initial interview.

Show Slide 46

3. Follow interviewing and documentation procedures in subsections C, Conducting the Interview, and D, Recording Witness Recollections.

   👁 Go back and review this material. (See pages 15–22. Refer students to *Guide* pages 22–24.)

4. Provide no information from other sources.

   👁 Witnesses may ask the investigator about information that has developed since the initial interview. Providing the witness with specific information obtained from other witnesses or from physical evidence may influence the witness's perception of the incident.

   👁 Should other information arise following the initial interview that differs from, contradicts, or corroborates information the witness provided, this information can be clarified with the witness at this time. However, the investigator can present that information to the witness in a nonleading manner. The investigator can provide the witness with neutral information, such as asking if any vehicle was present at the time of the incident, NOT "Are you sure there was not a blue Ford at the scene?"

Show Slide 47

Summary: Reestablishing contact and rapport with the witness often leads to recovery of additional information. Maintaining open communication channels with the witness throughout the investigation can lead to additional evidence.

Sample Lesson Plan: Identification

Section II. Mug Books and Composites

Show Slides 48–50

A. Preparing Mug Books

This subsection covers photo mug books and displays that use computerized imaging systems.

NOTE: "Mug books" (i.e., collections of photos of previously arrested persons) may be used in cases in which a suspect has not yet been determined and other reliable sources have been exhausted. This technique may provide investigative leads, but results should be evaluated with caution.

Principle: Nonsuggestive composition of a mug book may enable the witness to provide a lead in a case in which no suspect has been determined and other reliable sources have been exhausted.

Policy: The investigator/mug book preparer should compose the mug book in such a manner that individual photos are not suggestive.

Procedure: In selecting photos to be preserved in a mug book, the preparer should—

1. Group photos by format (e.g., color or black and white; Polaroid, 35mm, or digital; video) to ensure that no photo unduly stands out.

   👁 All photos should be the same format so that no individual photo stands out to a witness. For example, one color photo shown among a group of black-and-white photos might suggest to a witness that the color photo is of a more recent offender and, therefore, more likely to be the perpetrator of a recent crime. Also, different photo formats show varying levels of detail.

2. Select photos of individuals that are uniform with regard to general physical characteristics (e.g., race, age, sex).

Show Slide 51

   👁 A witness will usually have an idea of a perpetrator's general physical characteristics, so sorting mug books by race, age, or sex can facilitate the witness's task (i.e., the witness will not need to look through photos of young black females when the perpetrator was described as a middle-aged white male).

Show Slide 52

3. Consider grouping photos by specific crime (e.g., sexual assault, gang activity).

   👁 This can also facilitate the witness's task. For example, sex offenders tend to be recidivists, so a collection of photos of sex offenders may be useful to a witness/victim of a sexual assault.

4. Ensure that positive identifying information exists for all individuals portrayed.

   👁 If a witness selects a photo, identifying information will be needed for subsequent investigation, departmental records, and/or to provide the information for court purposes.

5. Use reasonably contemporary photos.

   👁 This is necessary because appearances change over time.

IMPORTANT:
Emphasize that the purpose of this step is to minimize the suggestiveness of the procedure.

6. Use only one photo of each individual in the mug book.

   👁 The presence of more than one photo of an individual in a mug book increases the chances of that individual being selected by a witness, thereby increasing the suggestiveness of the procedure.

Show Slide 53

Summary: Mug books should be objectively compiled to yield investigative leads that will be admissible in court.

Show Slide 54

B. Developing and Using Composite Images

NOTE: Composite images can be beneficial investigative tools. However, they are rarely used as stand-alone evidence.

Principle: Composites provide a depiction that may be used to develop investigative leads.

Policy: The person preparing the composite should select and employ the composite technique in such a manner that the witness's description is reasonably depicted.

Procedure: The person preparing the composite should—

1. Assess the ability of the witness to provide a description of the perpetrator.

   👁 Assess the physical and mental state of the witness at both the time of the procedure and the time of the incident to determine if any conditions are or were present that could interfere with the witness's ability to give an adequate description of the perpetrator.

2. Select the procedure to be used from those available (e.g., identikit-type, artist, or computer-generated images).

👁 This choice may be based on the equipment, training, and experience available in each department or jurisdiction.

3. Unless part of the procedure, avoid showing the witness any photos immediately prior to development of the composite.

👁 Showing photos to the witness immediately prior to the procedure could influence the description he/she provides.

Show Slide 55

4. Select an environment for conducting the procedure that minimizes distractions.

👁 This will enable the witness to concentrate and provide a more detailed and complete description.

5. Conduct the procedure with each witness separately.

Show Slide 56

👁 Witnesses must be separated so they are not influenced by descriptions others provide.

6. Determine with the witness whether the composite is a reasonable representation of the perpetrator.

👁 Allowing the witness to view the completed composite gives the witness an opportunity to suggest changes and may thereby produce a better likeness of the perpetrator. It also allows the witness to state whether the image is a reasonable likeness of the perpetrator. For example, the witness can be asked to rate the image as to its accuracy and/or its potential usefulness.

Summary: The use of composite images can yield investigative leads in cases in which no suspect has been determined. Use of these procedures can facilitate obtaining from the witness a description that will enable the development of a reasonable likeness of the perpetrator.

Show Slide 57

C. Instructing the Witness

Show Slide 58

Principle: Instructions to the witness prior to conducting the procedure can facilitate the witness's recollection of the perpetrator.

Policy: The investigator/person conducting the procedure should provide instructions to the witness prior to conducting the procedure.

Procedure:

Mug Book: The investigator/person conducting the procedure should—

1. Instruct each witness without other persons present.

   👁 This minimizes distractions and allows the witness to concentrate.

2. Describe the mug book to the witness only as a "collection of photographs."

   👁 The witness should not be told anything that might influence his/her decision to choose a photo, such as the fact that the individuals portrayed have arrest records, the offenses for which the individuals were arrested, or the geographical area with which they are associated.

Show Slide 59

IMPORTANT:

Emphasize that the witness should not feel pressured to select a photo.

3. Instruct the witness that the person who committed the crime may or may not be present in the mug book.

   👁 This is important so that the witness will not feel pressured to make a selection even if none of the photos resemble the perpetrator. This also will help to prevent a misidentification.

4. Consider suggesting to the witness to think back to the event and his/her frame of mind at the time.

   👁 Recreating the circumstances of the event makes memory more accessible. Instruct the witness to think about his/her thoughts and feelings at the time of the incident.

Show Slide 60

5. Instruct the witness to select a photograph if he/she can and to state how he/she knows the person if he/she can.

   👁 Witnesses may recognize a photo for reasons other than it being a photo of the perpetrator. Therefore, it is important to determine how or from where the witness knows the depicted individual. For example, the witness may recognize someone he/she just saw in the precinct lobby.

6. Assure the witness that regardless of whether he/she makes an identification, the police will continue to investigate the case.

   👁 This will help the witness to relax and help to alleviate any pressure the witness may feel to make a selection.

Show Slide 61

7. Instruct the witness that the procedure requires the investigator to ask the witness to state, in his/her own words, how certain he/she is of any identification.

👁 It can be helpful to have some indication of how certain the witness is at the time of the identification. This can be useful in assessing the likelihood of whether or not the identification is accurate. Later, the witness's certainty might be influenced by other factors.

👁 It is not necessary for the witness to give a number to express his/her certainty. Some witnesses will spontaneously include information about certainty (e.g., "That's him, I KNOW that's him," or, "It could be that one"). If the witness does not volunteer information about certainty, then the witness can be asked to state certainty in his/her own words. A question such as, "How do you know this individual?" will often lead the witness to express his/her certainty. If a statement of certainty is not obtained, then the investigator can follow up with the question, "How certain are you?"

NOTE: If a witness selects a photo from the mug book, using that same photo in a later identification procedure with that same witness can lead to challenges to that procedure. Using a different or more recent photo in a followup procedure may be acceptable.

IMPORTANT:
Emphasize the importance of recording a certainty statement.

Composite: The investigator/person conducting the procedure should—

Show Slide 62

1. Instruct each witness without other persons present.

   👁 This minimizes distractions and allows the witness to concentrate.

2. Explain the type of composite technique to be used.

   👁 The witness needs to understand what will be required of him/her.

3. Explain to the witness how the composite will be used in the investigation.

   Show Slide 63

   👁 This will help the witness understand that the purpose of the composite is to develop investigative leads.

4. Instruct the witness to think back to the event and his/her frame of mind at the time.

   👁 Recreating the circumstances of the event makes memory more accessible. Instruct the witness to think about his/her thoughts and feelings at the time of the incident.

Summary: Providing instructions to the witness can improve his/her comfort level and can result in information that may assist the investigation.

Show Slide 64

Show Slide 65

NOTE:
These procedures should be reviewed, however an elaborate explanation is not necessary.

Show Slide 66

Show Slide 67

Show Slides 68–69

IMPORTANT:
Discuss with the class the inherent suggestiveness of this procedure.

D. Documenting the Procedure

Principle: Documentation of the procedure provides an accurate record of the results obtained from the witness.

Policy: The person conducting the procedure should preserve the outcome of the procedure by accurately documenting the type of procedure(s) employed and the results.

Procedure: The person conducting the procedure should—

1. Document the procedure employed (e.g., identikit-type, mug book, artist, computer-generated image) in writing.

2. Document the results of the procedure in writing, including the witness's own words regarding how certain he/she is of any identification.

3. Document items used and preserve composites generated.

Summary: Documentation of the procedure and its outcome can be an important factor in the investigation and any subsequent court proceedings.

Section IV. Field Identification Procedure (Showup)

A. Conducting Showups

Principle: When circumstances require the prompt display of a single suspect to a witness, challenges to the inherent suggestiveness of the encounter can be minimized through the use of procedural safeguards.

Policy: The investigator should use procedures that avoid unnecessary suggestiveness.

Procedure: When conducting a showup, the investigator should—

1. Determine and document, prior to the showup, a description of the perpetrator.

2. Consider transporting the witness to the location of the detained suspect to limit the legal impact of the suspect's detention.

   👁 There are likely to be legal restrictions concerning transporting suspects to the scene. Local/jurisdictional laws or policies should be consulted and followed. Other issues that may be involved with bringing the suspect to the scene include potential contamination of the scene or exposure to media or multiple witnesses.

3. When multiple witnesses are involved—

Show Slide 70

   a. Separate witnesses and request that they avoid discussing details of the incident with other witnesses.

   👁 Witnesses should not hear others' accounts because they may be influenced by that information.

   b. If a positive identification is obtained from one witness, consider using other identification procedures (e.g., lineup or photo array) for remaining witnesses.

   👁 Because showups can be considered inherently suggestive, once an identification is obtained at a showup and probable cause for arrest has been achieved, less suggestive procedures can be used with other witnesses to obtain their identifications.

4. Caution the witness that the person he/she is looking at may or may not be the perpetrator.

Show Slide 71

   IMPORTANT: Emphasize why this instruction is important.

   👁 This instruction to the witness can lessen the pressure on the witness to make an identification solely to please the investigator or because the witness feels it is his/her duty to do so. The investigator should assure the witness that the investigation will continue regardless of whether an identification is obtained at the showup. Keep in mind that it is just as important to clear innocent parties; a nonidentification can help to refocus the investigation.

5. Obtain and document a statement of certainty for both identifications and nonidentifications.

   👁 It can be helpful to have some indication of how certain the witness is at the time of an identification (or nonidentification). This can be useful in assessing the likelihood of whether or not the identification is accurate. Later, the witness's certainty might be influenced by other factors.

   👁 It is not necessary for the witness to give a number to express his/her certainty. Some witnesses will spontaneously include information about certainty (e.g., "That's him, I KNOW that's him," or, "It could be him"). If the witness does not volunteer information about certainty, then the witness can be asked to state certainty in his/her own words. A question such as, "How do you know this individual?" will often lead the witness to express his/her certainty. If a statement of certainty is not obtained, then the investigator can follow up with the question, "How certain are you?"

<table>
<tr>
<td>

Show Slide 72

</td>
<td>

Summary: The use of a showup can provide investigative information at an early stage, but the careful use of procedural safeguards can mitigate the inherent suggestiveness of a showup.

</td>
</tr>
<tr>
<td>

Show Slide 73

</td>
<td>

B. Recording Showup Results

Principle: The record of the outcome of the field identification procedure accurately and completely reflects the identification results obtained from the witness.

Policy: When conducting a showup, the investigator should preserve the outcome of the procedure by documenting any identification or nonidentification results obtained from the witness.

Procedure: When conducting a showup, the investigator should—

1. Document the time and location of the procedure.

2. Record both identification and nonidentification results in writing, including the witness's own words regarding how certain he/she is.

</td>
</tr>
<tr>
<td>

<u>NOTE:</u>
These procedures should be reviewed, however an explanation is generally unnecessary.

Show Slide 74

</td>
<td>

Summary: A complete and accurate record of the outcome of the showup can be a critical document in the investigation and any subsequent court proceedings.

</td>
</tr>
<tr>
<td>

Show Slide 75

Show Slide 76:
Play Video Clip 1:
Follow instructions to conduct exercise.

</td>
<td>

Section V. Procedures for Eyewitness Identification of Suspects

Before instructing section V, consider playing video clip 1. Only the incident video is shown at this point. Do not provide any instructions to the students prior to viewing the clip other than to watch the screen. The idea is to catch the students by surprise the way that most eyewitnesses are caught. Once they have viewed the clip, move on to the procedural instruction below (the lineup videos will be viewed later).

</td>
</tr>
<tr>
<td>

Show Slide 77

</td>
<td>

A. Composing Lineups

Principle: Fair composition of a lineup enables the witness to provide a more accurate identification or nonidentification.

Policy: The investigator should compose the lineup in such a manner that the suspect does not unduly stand out.

</td>
</tr>
</table>

Procedure:

Photo Lineup: In composing a photo lineup, the investigator should:

1. Include only one suspect in each identification procedure.

   👁 The problem with multiple-suspect lineups is that the probability of a possible mistaken identification rises dramatically as the number of suspects in a lineup increases. If more than one suspect must be shown in any one lineup, the fillers must be multiplied accordingly (e.g., 2 suspects require a minimum of 10 fillers).

2. Select fillers who generally fit the witness's description of the perpetrator. When there is a limited/inadequate description of the perpetrator provided by the witness, or when the description of the perpetrator differs significantly from the appearance of the suspect, fillers should resemble the suspect in significant features.

   👁 This does not mean that the fillers must closely resemble the suspect (see notes under procedure 5 below). If the description does not fit the suspect on some characteristic (e.g., the witness described dark hair, yet the suspect has light hair), then the fillers should match the suspect on that characteristic rather than matching the description on that characteristic so that the suspect does not unduly stand out.

3. If multiple photos of the suspect are reasonably available to the investigator, select a photo that resembles the suspect's description or appearance at the time of the incident.

   👁 The most recent photo of the suspect is not necessarily the best one to use if the suspect's appearance has changed since the time of the crime. For example, the suspect may intentionally change his/her appearance.

4. Include a *minimum* of five fillers (nonsuspects) per identification procedure.

   👁 This is a suggested minimum number; some jurisdictions might require more fillers.

5. Consider that complete uniformity of features is not required. Avoid using fillers that so closely resemble the suspect that a person familiar with the suspect might find it difficult to distinguish the suspect from the fillers.

   👁 In their efforts to ensure that the suspect's photo does not unduly stand out, police have often gone to great lengths to ensure that all members of a lineup look as similar to one another as possible,

---

IMPORTANT:
Clarify that this procedure assumes a case with only one perpetrator.

Show Slide 78

Show Slide 79:
EXERCISE:
Provide description of perpetrator and have students select appropriate fillers. (The best choices are 1, 3, 5, 7, and 10.)

Show Slide 80:
EXERCISE:
Show photo of suspect and have students select fillers based on suspect features. (The best choices are 1, 3, 8, 11, and either 4 or 10.)

Show Slide 81

Show Slide 82

Show Slide 83:

IMPORTANT:

Emphasize the difficulties of using fillers that are too similar. Consider conducting another filler-selection exercise to demonstrate this point.

Show Slide 84

Show Slide 85

Show Slide 86

EXERCISE:

Consider having a student administer separate photo lineups to two students. Did the administering student think to change the position of the suspect in the second lineup?

including the suspect. Making the fillers closely resemble the suspect is not advised because a lineup in which all the people look very similar to one another actually reduces the chances of an accurate identification by a witness. According to procedures 2, 5, 6, and 10, lineup fillers must merely match the *description* of the offender as given by the witness viewing that lineup, as long as the policy is upheld that the suspect does not unduly stand out.

6. Consider creating a consistent appearance between the suspect and fillers with respect to any unique or unusual feature (e.g., scars or tattoos) used to describe the perpetrator by artificially adding or concealing that feature.

   👁 If there is a unique feature/characteristic described by the witness, such as a scar, the preferred procedure is to leave the unique feature visible and select fillers with a similar feature/characteristic. Sometimes police choose to enhance the fillers with a similar feature (still ensuring that the suspect does not unduly stand out). If the suspect has a unique feature not described by the witness, you should not alter the suspect's photo. Rather you should select fillers that have a similar, but not identical, feature or enhance the fillers with a similar feature.

   👁 Slide 85 is a photo lineup from a case in which the witness described the perpetrator as being a cross-eyed black male. The investigator in this case was unable to find cross-eyed black males to serve as fillers, so he chose to create this photo lineup using imaging software on a computer to cross the eyes of the fillers.

7. Consider placing suspects in different positions in each lineup, both across cases and with multiple witnesses in the same case. Position the suspect randomly in the lineup.

   👁 If specific investigators consistently choose the same lineup location for the suspect, this can become common knowledge among both law enforcement officers and the general public. This could lead a witness to pick the person in that position for reasons other than recognizing the suspect.

   👁 Some witnesses can be reserved for alternative identification procedures, such as a live lineup or a different photo lineup. For example, your original identification procedure may be found to be inadmissible in court, whereas an alternative procedure (e.g., a live lineup) or a second photo lineup may be admissible.

8. When showing a new suspect, avoid reusing fillers in lineups shown to the same witness.

   👁 Using the same fillers with a new suspect can make the suspect stand out as the only one not appearing in a previous photo lineup. This could be considered a suggestive procedure. Also, the witness might recognize one of the fillers (from seeing him/her in a previous lineup) and misidentify the filler as the perpetrator.

9. Ensure that no writings or information concerning previous arrest(s) will be visible to the witness.

   👁 Some witnesses might try to extract meaning from any arrest dates or other markings on the photos. Such information could lead some witnesses to make faulty inferences. Booking plates, for instance, can be covered with tape. Also ensure that no writings indicating previous witnesses' identifications are visible to the witness.

Show Slide 87

10. View the spread, once completed, to ensure that the suspect does not unduly stand out.

    👁 Consider showing the photo lineup to people unfamiliar with the case and ask them if they can identify the suspect. In general, if the photo lineup is properly constructed, a person who is given the verbal description of the perpetrator (as described by the witness) should not be able to tell which person is the suspect in the case.

11. Preserve the presentation order of the photo lineup. In addition, the photos themselves should be preserved in their original condition.

    👁 In order to defend legal challenges to the lineup procedures, it is critical to reproduce the original lineup for presentation in future proceedings. It is advisable to retain the original photos as evidence or, alternatively, photocopy (in color if possible) the original lineup to produce a copy in the event that one or more of the original photographs cannot be reproduced and to preserve an accurate representation of the order of the photos.

Show Slide 88

**Show Slide 89:
EXERCISE:**
Have students critique lineup composition. (General problems: The fillers do not fit the witness's description of the perpetrator, nor do they match the suspect in significant features; the suspect stands out.)

Live Lineup:

Note how the criteria for selecting fillers for a photo lineup are the same as the criteria for selecting fillers for a live lineup (except for the minimum number of fillers).

In composing a live lineup, the investigator should—

1. Include only one suspect in each identification procedure.

   👁 In multiple-suspect lineups, the probability of a possible mistaken identification rises as the number of suspects in a lineup increases. If more than one suspect must be presented in any one lineup, the

Show Slide 90

Much of the information in this subsection is substantially the same as that covered for photo lineups, so only a cursory review is needed.

fillers should be multiplied accordingly (e.g., two suspects indicate a minimum of eight fillers).

Show Slide 91

2. **Select fillers who generally fit the witness's description of the perpetrator. When there is a limited/inadequate description of the perpetrator provided by the witness, or when the description of the perpetrator differs significantly from the appearance of the suspect, fillers should resemble the suspect in significant features.**

   👁 This does not mean that the fillers must closely resemble the suspect (see notes under procedure 6 below). If the description does not fit the suspect on some characteristic (e.g., the witness described dark hair, yet the suspect has light hair), then the fillers should match the suspect on that characteristic rather than matching the description on that characteristic so that the suspect does not stand out.

Show Slide 92

3. **Consider placing suspects in different positions in each lineup, both across cases and with multiple witnesses in the same case. Position the suspect randomly, unless, where local practice allows, the suspect or the suspect's attorney requests a particular position.**

   👁 If specific investigators consistently choose the same lineup location for the suspect, this can become common knowledge among both law enforcement officers and the general public. This could lead a witness to pick the person in that position for reasons other than recognizing the suspect.

   👁 Some witnesses can be reserved for alternative identification procedures, such as a photo lineup or a different live lineup. For example, your original identification procedure may be found to be inadmissible in court, whereas an alternative procedure (e.g., a photo lineup) or a second live lineup may be admissible.

Show Slide 93:
IMPORTANT:
Emphasize that the minimum number of fillers (four) for a live lineup is different than for a photo lineup.

4. **Include a *minimum* of four fillers (nonsuspects) per identification procedure.**

   👁 The fact that a fewer number of fillers is required for a live lineup than for a photo lineup is purely a practical consideration. This is a suggested minimum. It is more difficult to obtain people to use as fillers in a live lineup than it is to obtain photos to use as fillers for a photo lineup.

5. **When showing a new suspect, avoid reusing fillers in lineups shown to the same witness.**

   👁 Using the same fillers with a new suspect can make the suspect stand out as the only one not appearing in a previous lineup. This could be considered a suggestive procedure. Also, the witness might recognize one of the fillers (from seeing him/her in a previous lineup) and misidentify the filler as the perpetrator.

6. **Consider that complete uniformity of features is not required. Avoid using fillers that so closely resemble the suspect that a person familiar with the suspect might find it difficult to distinguish the suspect from the fillers.**

   👁 In their efforts to ensure that the suspect does not unduly stand out, police have often gone to great lengths to ensure that all members of a lineup look as similar to one another as possible, including the suspect. Selecting fillers that closely resemble the suspect is not advised because a lineup in which all the people look very similar to one another actually reduces the chances of an accurate identification by a witness. According to procedures 2, 6, and 7, lineup fillers must merely match the *description* of the offender as given by the witness viewing that lineup, as long as the policy is upheld that the suspect does not unduly stand out.

   Show Slide 94

7. **Consider creating a consistent appearance between the suspect and fillers with respect to any unique or unusual feature (e.g., scars, tattoos) used to describe the perpetrator by artificially adding or concealing that feature.**

   Show Slide 95

   👁 If there is a unique feature/characteristic described by the witness, such as a scar, police sometimes choose to leave the unique feature visible and select fillers with a similar feature/characteristic or enhance the fillers with a similar feature (still ensuring that the suspect does not unduly stand out). If the suspect has a unique feature not described by the witness, you should not alter the suspect's appearance. Rather you should select fillers that have a similar, but not identical, feature or enhance the fillers with a similar feature.

   Show Slide 96: **EXERCISE:** Ask the students to evaluate the adequacy of the lineup. (Two problems: Too few fillers are included, and number 2 stands out as the only participant with light-colored hair.)

Summary: These suggestions can help produce a lineup in which the suspect does not unduly stand out. An identification obtained through a lineup composed in this manner may have stronger evidentiary value.

Show Slide 97

Now show the video clips of the live lineups to complete the exercise begun at the start of this section.* Most students will pick someone from the video lineup and will be surprised when you tell them that the actual perpetrator is not in the lineup. Play the video of the event again so that the students can see the actual perpetrator and note how he is not simply a "lookalike" for those in the lineup.** Explain to them at this point that the most difficult problem that witnesses confront in a lineup is when the actual perpetrator is not in the lineup.

Show Slides 98–99: Play Video Clips 2 and 3: Follow instructions to complete exercise.

Show Slide 100: Replay Video Clip 1

Explain to the students how eyewitnesses have natural tendencies to select someone from a lineup who looks most like the perpetrator *relative* to the other lineup members. Although this strategy works well if the perpetrator is in the lineup, there are times when the actual perpetrator is not in the lineup.

Lead a class discussion of the video exercise.

Explain to the students that the suggestions described in the *Guide* for conducting photographic and live lineups are designed to minimize the chances of mistaken identification while still permitting witnesses to identify the actual perpetrator. Point out that the lineup used in the video was a poor example of how a lineup should be constructed and that the viewing instructions given were poor (only one suspect fits the original description and instructions failed to indicate that the perpetrator may or may not be in the lineup).

Show Slide 101

B. Instructing the Witness Prior to Viewing a Lineup

Discuss the problem of "relative judgments."

Much of the material in this section should help prevent the witness from making "relative judgments." Relative judgments occur when witnesses encounter a lineup in which the actual perpetrator is not in the lineup (i.e., the suspect is not the actual perpetrator). Research shows that eyewitnesses tend to select the person who looks most like the perpetrator relative to the other lineup members. The fact that police are showing a lineup to a witness can lead some witnesses to presume that the actual perpetrator will be in the lineup. These instructions are designed to help reduce the tendency for witnesses to make this assumption.

Principle: Instructions given to the witness prior to viewing can facilitate an identification or nonidentification based on his/her own memory.

Policy: Prior to presenting a lineup, the investigator should provide instructions to the witness to ensure the witness understands that the purpose of the identification procedure is to exculpate the innocent as well as to identify the actual perpetrator.

Procedure:

Photo Lineup: Prior to presenting a photo lineup, the investigator should—

1. Advise the witness that he/she will be asked to view a set of photographs.

2. Advise the witness that it is just as important to clear innocent persons from suspicion as to identify guilty parties.

   👁 Because the suspect in the case might *not* be the actual offender, the identification procedure can in fact help clear innocent persons from suspicion. This instruction helps emphasize that failure to identify the suspect might be, in some cases, the

appropriate outcome. Clearing an innocent suspect from suspicion can help refocus the investigation on developing other suspects.	
3. Advise the witness that individuals depicted in lineup photos may not appear exactly as they did on the date of the incident because features such as head and facial hair are subject to change.	**Show Slide 102**
👁 Many physical characteristics are changeable. Hair, for instance, can be restyled, colored, cut, or grown longer; facial hair can be grown or cut; and so forth. Witnesses need to keep in mind that the suspect's appearance on these changeable features might have been different at the time of the photo than it was at the time of the crime.	
4. Advise the witness that the person who committed the crime may or may not be in the set of photographs being presented.	**Show Slide 103**
👁 This training seeks to prevent the misidentification of an innocent suspect. It is important to emphasize that the person who committed the crime may not be present. It does not weaken the investigation if the actual perpetrator is not in the lineup and the witness does not make a selection. In fact, it may benefit the investigation by strengthening the witness's credibility and helping to refocus the investigation.	
5. Assure the witness that regardless of whether an identification is made, the police will continue to investigate the incident.	
👁 This instruction lessens the pressure on the witness to make an identification and reassures the witness that the progress of the investigation does not hinge solely on his/her identification. Even if the witness does not make an identification, the investigation should continue.	
6. When appropriate, advise the witness that the procedure requires the investigator to ask the witness to state, in his/her own words, how certain he/she is of any identification.	**Show Slide 104**
👁 It can be helpful to have some indication of how certain the witness is at the time of the identification. This can be useful in assessing the likelihood of whether or not the identification is accurate. Later, the witness's certainty might be influenced by other factors.	
👁 It is not necessary for the witness to give a number to express his/her certainty. Some witnesses will spontaneously include information about certainty (e.g., "That's him, I KNOW that's	

him," or "It could be number three."). If the witness does not volunteer information about certainty, then the witness can be asked to state certainty in his/her own words. A question such as, "How do you know this individual?" will often lead the witness to express his/her certainty. If a statement of certainty is not obtained, then the investigator can follow up with the question, "How certain are you?"

Show Slide 105
The information in this subsection is substantially the same as that covered for photo lineups, so only a cursory review is needed.

Live Lineup: Prior to presenting a live lineup, the investigator should—

1. Advise the witness that he/she will be asked to view a group of individuals.

2. Advise the witness that it is just as important to clear innocent persons from suspicion as to identify guilty parties.

   ◉ Because the suspect in the case might *not* be the actual offender, the identification procedure can in fact help clear innocent persons from suspicion. This advice helps emphasize that failure to identify the suspect might be, in some cases, the appropriate outcome. Clearing an innocent suspect from suspicion can help refocus the investigation on developing other suspects.

Show Slide 106

3. Advise the witness that individuals present in the lineup may not appear exactly as they did on the date of the incident, as features such as head and facial hair are subject to change.

   ◉ Many physical characteristics are changeable. Hair, for instance, can be restyled, colored, cut, grown longer; facial hair can be grown or cut; and so forth. Witnesses need to keep in mind that the suspect's appearance on these changeable features might be different at the time of the lineup than it was at the time of the crime.

4. Advise the witness that the person who committed the crime may or may not be present in the group of individuals.

   ◉ This training seeks to prevent the misidentification of an innocent suspect. It is important to emphasize that the person who committed the crime may not be present. It does not weaken the investigation if the actual perpetrator is not in the lineup and the witness does not make a selection. In fact, it may benefit the investigation by strengthening the witness's credibility and helping to refocus the investigation.

Show Slide 107

5. Assure the witness that regardless of whether an identification is made, the police will continue to investigate the incident.

   ◉ This lessens the pressure on the witness to make an identification and reassures the witness that the progress of the investigation does not hinge solely on his/her identification. Even if the witness does not make an identification, the investigation will continue.

6. When appropriate, advise the witness that the procedure requires the investigator to ask the witness to state, in his/her own words, how certain he/she is of any identification.

👁 It can be helpful to have some indication of how certain the witness is at the time of the identification. It can be useful in assessing the likelihood of whether or not the identification is accurate. Later, the witness's certainty might be influenced by other factors.

👁 It is not necessary for the witness to give a number to express his/her certainty. Some witnesses will spontaneously include information about certainty (e.g., "That's him, I KNOW that's him," or "It could be number 3."). If the witness does not volunteer information about certainty, then the witness should be asked to state certainty in his/her own words. A question such as, "How do you know this individual?" will often lead the witness to express his/her certainty. If a statement of certainty is not obtained, then the investigator should follow up with the question, "How certain are you?"

Summary: Appropriate information provided to the witness prior to presentation of a lineup will likely improve the accuracy and reliability of any identification obtained from the witness and can facilitate the elimination of innocent parties from the investigation.

C. Conducting the Identification Procedure

Explain to students the distinction between a simultaneous and a sequential identification procedure. In a simultaneous identification procedure, all members of the lineup are shown to the witness at the same time. This allows the witness to compare all lineup members before making a decision. In a sequential lineup procedure, however, the witness views only one member of the lineup at a time. The witness must make a decision on each lineup member before viewing the next lineup member.

A major difference between the simultaneous and sequential procedure is that the sequential procedure tends to prevent the eyewitness from making relative judgments. Recall that relative judgments can be problematic because they involve comparing one lineup member to another and picking the person who most looks like the perpetrator. The sequential procedure leads witnesses to compare each lineup member with their memory of the perpetrator rather than comparing one lineup member with another lineup member. Relative judgments can also be

Show Slide 108
<u>EXERCISE:</u>
Have students give each other mock lineup viewing instructions.

Show Slide 109
<u>NOTE:</u>
Discuss the distinction between *simultaneous* and *sequential* lineup procedures, including examples of the merits of the sequential lineup.

reduced even with a simultaneous procedure by using suggestions on composing, instructing witnesses on, and conducting simultaneous lineups described in the *Guide*.

Some jurisdictions may want to consider using "blind" identification procedures. In a typical blind identification procedure, the person who conducts the lineup does not know which person in the lineup is the suspect. Using this type of procedure, the case investigator simply has someone conduct the lineup who is not familiar with the case, not familiar with the identity of the lineup members, and does not know the lineup position of the suspect. Such a procedure helps ensure not only that the case investigator does not unintentionally influence the witness but also that there can be no arguments later (e.g., at trial) that the witness's selection or statements at the lineup were influenced by the case investigator.

Although an awareness on the part of the investigator that he/she should do nothing to influence the witness's choice or certainty can be sufficient to ensure that such influence does not occur, some jurisdictions might nevertheless prefer to use blind testing techniques. Students can be told about blind identification procedures in the context of discussions about how to avoid influencing the witness.

Principle: The identification procedure should be conducted in a manner that promotes the reliability, fairness, and objectivity of the witness's identification.

Policy: The investigator should conduct the lineup in a manner conducive to obtaining accurate identification or nonidentification decisions.

Procedure:

Simultaneous Photo Lineup:
When presenting a simultaneous photo lineup, the investigator should—

1. Provide viewing information to the witness as outlined in subsection B, Instructing the Witness Prior to Viewing a Lineup.

2. Confirm that the witness understands the nature of the lineup procedure.

   👁 Investigators should make sure that the witness understands everything at this point. For example, witnesses can be asked, "Do you understand?" or "Do you have any questions?"

3. Avoid saying anything to the witness that may influence the witness's selection.

   👁 Ideally, nothing should be said to the witness because it might indicate which person the investigator believes is the perpetrator

Discuss the merits of "blind" procedures.

NOTE:
Much of the procedural information in this subsection is repetitive and need only be explained once, then reviewed as needed.

Show Slide 110

or that the investigator believes the perpetrator is definitely in the lineup. Also, anything said to the witness might interfere with his/her ability to concentrate on the task. If something needs to be said to facilitate the procedure, it must not convey any information about the identity of the suspect (e.g., NOT "I noticed you pointed at number two," BUT rather "Would it help for me to explain the instructions again?").

4. If an identification is made, avoid reporting to the witness any information regarding the individual he/she has selected prior to obtaining the witness's statement of certainty.

   👁 The witness should not be told anything about the status of the person identified at this point (e.g., do not say anything that validates the witness's selection, such as, "That's the person we have as a suspect," or "That's the same person that another witness picked"; do not say anything that discounts the witness's selection, such as, "That person is not a suspect"). This includes nonverbal reactions, such as facial expressions of approval or disapproval. Such reactions can influence the certainty (confidence level) that the witness expresses in his/her choice.

   👁 A witness may identify a suspect from a lineup and the investigators later uncover evidence clearing that suspect. Inadvertently reinforcing the witness's selection (e.g., "That was our suspect") will make it difficult to show that witness another lineup with a new suspect. It can be acceptable to share the results of the identification at a later time, but not before the witness's level of certainty has been ascertained.

5. Record any identification results and witness's statement of certainty as outlined in subsection D, Recording Identification Results. | Show Slide 111

   👁 Some departments have a form on which to record the results of a lineup identification procedure. Usually, such forms have a place to enter the number of the lineup member who was selected (if any), the name and other identifying information of the witness, the date the lineup was held, the name of the investigator who administered the lineup and the names of others who might have been present, a case number, and lines for the signatures of the witness and the investigator. This form may also include space for the witness to write out a statement about the identification. | Show Slide 112

6. Document in writing the photo lineup procedures, including— | Show Slide 113

   a. Identification information and sources of all photos used.

   b. Names of all persons present at the photo lineup.

   c. Date and time of the identification procedure.

Show Slide 114	7. Advise the witness not to discuss the identification procedure or its results with other witnesses involved in the case and discourage contact with the media.

👁 Remind the witness that discussing the results of the procedure could harm the investigation. Such discussion by the witness may influence other witnesses' identification decisions or their certainty.

👁 Witnesses can be warned at this time that the positioning of the lineup members might be changed for other witnesses and that it is important not to try to influence another witness. It is important that witnesses reach decisions independently, not only for investigative purposes but also for later proceedings.

Show Slide 115

Sequential Photo Lineup:

The sequential procedure is quite different from the simultaneous procedure. The sequential decision procedure is meant to reduce the tendency of the witness to compare one photo with another photo (i.e., make relative judgments). The idea is for the witness to make a final decision on each photo before moving on to the next photo by comparing each photo with his/her memory of the perpetrator.

When presenting a sequential photo lineup, the investigator should—

1. Provide viewing information to the witness as outlined in subsection B, Instructing the Witness Prior to Viewing a Lineup.

Show Slide 116

Demonstrate to the class how to conduct a sequential photo lineup procedure.

2. Provide the following *additional* viewing information to the witness:

a. Individual photographs will be viewed *one at a time.*

b. The photos are in random order.

c. Take as much time as needed in making a decision about each photo before moving on to the next one.

Show Slide 117

d. All photos will be shown, even if an identification is made; ***or*** the procedure will be stopped at the point of an identification (consistent with jurisdictional/departmental procedures).

👁 The investigator should follow a fixed technique as to whether the procedure will stop when the witness makes a selection of a photo or whether the procedure will continue until all photos are presented. If the investigator sometimes continues to show photos and sometimes does not, it could appear that the decision to continue is being based on whether the witness is making the "right" pick.

3. Confirm that the witness understands the nature of the sequential procedure.

> 👁 Investigators should make sure that the witness understands everything at this point. Witnesses can be asked, "Do you understand?" or "Do you have any questions?"

4. Present each photo to the witness separately, in a previously determined order, removing those previously shown.

Show Slide 118

> 👁 Let the witness determine when to view the next photo (within a reasonable amount of time). There should not be more than one photograph displayed at once.

5. Avoid saying anything to the witness that may influence the witness's selection.

> 👁 Ideally, nothing should be said to the witness because it might indicate which person the investigator believes is the perpetrator or that the investigator believes that the perpetrator is definitely in the lineup. Also, anything said to the witness might interfere with his/her ability to concentrate on the task. If something needs to be said to facilitate the procedure, it should not convey any information about the identity of the suspect (e.g., NOT "I noticed you pointed at number two," BUT rather, "Would it help for me to explain the instructions again?"). Following this procedure is especially important with the sequential lineup because only one photo is being viewed at any given time.

6. If an identification is made, avoid reporting to the witness any information regarding the individual he/she has selected prior to obtaining any witness's statement of certainty.

Show Slide 119

> 👁 If the investigator wants to question the witness about certainty, the witness should not be told anything about the status of the person identified at this point (e.g., do not say, "That is the person we have as a suspect," or "That is the same person that another witness picked"; do not say anything that discounts the witness's selection, such as, "That person is not a suspect"). This includes nonverbal reactions, such as facial expressions of approval or disapproval. Such reactions could influence the certainty (confidence level) that the witness expresses in his/her choice.

> 👁 To make this more clear, consider the fact that a witness may identify a suspect from a lineup and the investigators later uncover evidence clearing that suspect. Inadvertently reinforcing the witness's selection (e.g., "That was our suspect") will make it difficult to show that witness another lineup with a new suspect. It

can be acceptable to share the results of the identification at a later time, but not before the witness's level of certainty has been ascertained.

Show Slide 120

7. Record any identification results and witness's statement of certainty as outlined in subsection D, Recording Identification Results.

Show Slide 121

8. Document in writing the photo lineup procedures, including—

   a. Identification information and sources of all photos used.

   b. Names of all persons present at the photo lineup.

   c. Date and time of the identification procedure.

Show Slide 122

9. Advise the witness not to discuss the identification procedure or its results with other witnesses involved in the case and discourage contact with the media.

   👁 Remind the witness that discussing the results of the procedure could harm the investigation. Such discussion by the witness may influence any other witnesses' identification decisions or their certainty.

   👁 Witnesses can be advised at this time that the positioning of the lineup members might be changed for other witnesses and that it is important not to try to influence another witness. Witnesses should reach decisions independently in order to aid the investigation and later proceedings.

EXERCISE:
Administer a photo lineup to a student in the class improperly (e.g., direct attention to a particular photo) and have students critique the error.

Show Slide 123

Simultaneous Live Lineup:

When presenting a simultaneous live lineup, the investigator/lineup administrator should—

1. Provide viewing information to the witness as outlined in subsection B, Instructing the Witness Prior to Viewing a Lineup.

Show Slide 124

2. Advise all those present at the lineup not to suggest in any way the position or identity of the suspect in the lineup.

3. Ensure that any identification actions (e.g., speaking, moving) are performed by all members of the lineup.

   👁 Even if the witness asks for only one person to walk or speak, all lineup members should be asked to perform the same action. Start with lineup member number one (as previously determined) and have each lineup member perform the action in order. (Consider that certain jurisdictions may have restrictions on what can be said by any lineup participant.)

4. Avoid saying anything to the witness that may influence the witness's selection.

👁 Ideally, nothing should be said to the witness at this point because it might indicate which person the investigator believes is the perpetrator or that the investigator believes the perpetrator is definitely in the lineup. Also, anything said to the witness might interfere with his/her ability to concentrate on the task. If something needs to be said to facilitate the procedure, it must not convey any information about the identity of the suspect (e.g., NOT "I noticed you pointed at number two," BUT rather "Would it help for me to explain the instructions again?").

5. If an identification is made, avoid reporting to the witness any information regarding the individual he/she has selected prior to obtaining any witness's statement of certainty.

👁 If the investigator wants to question the witness about certainty, the witness should not be told anything about the status of the person identified at this point (e.g., do not say, "That's the person we have as a suspect," or "That is the same person that another witness picked"; do not say, "That person is not a suspect"). This includes nonverbal reactions, such as facial expressions of approval or disapproval. Such reactions could influence the certainty (confidence level) that the witness expresses in his/her choice.

👁 To make this clearer, consider the fact that a witness may identify a suspect from a lineup and the investigators later uncover evidence clearing that suspect. Inadvertently reinforcing the witness's selection (e.g., "That was our suspect") will make it difficult to show that witness another lineup with a new suspect. It can be acceptable to share the results of the identification at a later time, but not before the witness's level of certainty has been ascertained.

6. Record any identification results and witness's statement of certainty as outlined in subsection D, Recording Identification Results.

7. Document the lineup in writing, including—

   a. Identification information of lineup participants.

   b. Names of all persons present at the lineup.

   c. Date and time the identification procedure was conducted.

8. Document the lineup by photo or video. This documentation should be of a quality that represents the lineup clearly and fairly.

	Show Slide 125
	Show Slide 126
	Show Slide 127
	Show Slide 128

9. Advise the witness not to discuss the identification procedure or its results with other witnesses involved in the case and discourage contact with the media.

👁 Remind the witness that discussing the results of the procedure could harm the investigation. Such discussion by the witness may influence any other witnesses' identification decisions or their certainty.

👁 Witnesses can be advised at this time that the positioning of the lineup members might be changed for other witnesses and that it is important not to try to influence another witness. It is important that witnesses reach decisions independently, not only for investigative purposes but also for later proceedings.

Show Slide 129

Sequential Live Lineup:

When presenting a sequential live lineup, the lineup administrator/ investigator should—

1. Provide viewing information to the witness as outlined in subsection B, Instructing the Witness Prior to Viewing a Lineup.

Show Slide 130

2. Provide the following *additional* viewing information to the witness:

   a. Individuals will be viewed *one at a time*.

Demonstrate to the class how to conduct a sequential live lineup.

   b. The individuals will be presented in random order.

   c. Take as much time as needed in making a decision about each individual before moving to the next one.

Show Slide 131

   d. If the person who committed the crime is present, identify him/her.

   e. All individuals will be presented, even if an identification is made; ***or*** the procedure will be stopped at the point of an identification (consistent with jurisdictional/departmental procedures).

   👁 The investigator should follow a fixed technique as to whether the procedure will stop when the witness makes a selection or whether the procedure will continue until all individuals are presented. If the investigator sometimes continues to show individuals and sometimes does not, it could appear that the decision to continue is being based on whether the witness is making the "right" pick.

3. Begin with all lineup participants out of the view of the witness.

Show Slide 132

4. Advise all those present at the lineup not to suggest in any way the position or identity of the suspect in the lineup.

5. Present each individual to the witness separately, in a previously determined order, removing those previously shown.

   👁 Let the witness determine when to view the next individual (within a reasonable amount of time). There should never be more than one individual displayed at once.

   👁 If the witness asks to view a particular lineup member again following the procedure, allow him/her to do so and document that fact.

6. Ensure that any identification actions (e.g., speaking, moving) are performed by all members of the lineup.   **Show Slide 133**

   👁 Even if the witness asks for only one person to walk or speak, all lineup members should be asked to perform the same action. Have each lineup member perform the action when they are presented. (Consider that certain jurisdictions may have restrictions on what can be said by any lineup participant.)

7. Avoid saying anything to the witness that may influence the witness's selection.

   👁 Ideally, nothing should be said to the witness because it might indicate which person the investigator believes is the perpetrator or that the investigator believes the perpetrator is definitely in the lineup. Also, anything said to the witness might interfere with his/her ability to concentrate on the task. If something needs to be said to facilitate the procedure, it must not convey any information about the identity of the suspect (e.g., NOT "I noticed you pointed at number two," BUT rather "Would it help for me to explain the instructions again?"). Following this procedure is especially important with the sequential lineup because only one individual is being viewed at any given time.

8. If an identification is made, avoid reporting to the witness any information regarding the individual he/she has selected prior to obtaining any witness's statement of certainty.   **Show Slide 134**

   👁 If the investigator wants to question the witness about certainty, the witness should not be told anything about the status of the person identified at this point (e.g., do not say, "That's the person we have as a suspect," or "That is the same person that another witness picked"; do not say, "That person is not a suspect"). This includes nonverbal reactions, such as facial expressions of approval or disapproval. Such reactions could influence the certainty (confidence level) that the witness expresses in his/her choice.

👁 To make this clearer, consider the fact that a witness may identify a suspect from a lineup and the investigators later uncover evidence clearing that suspect. Inadvertently reinforcing the witness's selection (e.g., "That was our suspect") will make it difficult to show that witness another lineup with a new suspect. It can be acceptable to share the results of the identification at a later time, but not before the witness's level of certainty has been ascertained.

9. Record any identification results and witness's statement of certainty as outlined in subsection D, Recording Identification Results.

Show Slide 135

10. Document the lineup procedures and content in writing, including—

   a. Identification information of lineup participants.

   b. Names of all persons present at the lineup.

   c. Date and time the identification procedure was conducted.

Show Slide 136

11. Document the lineup by photo or video. This documentation should be of a quality that represents the lineup clearly and fairly. Photo documentation can be of either the group or each individual.

Show Slide 137

12. Advise the witness not to discuss the identification procedure or its results with other witnesses involved in the case and discourage contact with the media.

   👁 Remind the witness that discussing the results of the procedure could harm the investigation. Such discussion by the witness may influence any other witnesses' identification decisions or their certainty.

   👁 Witnesses can be advised at this time that the positioning of the lineup members might be changed for other witnesses and that it is important to not try to influence another witness. It is important that eyewitnesses reach their decisions independently, not only for investigative purposes but also for later proceedings.

Show Slide 138

<u>**EXERCISE:**</u>
Administer a live lineup to the class improperly (e.g., do not advise the class that the perpetrator may not be present) and have students critique the error.

Summary: The manner in which an identification procedure is conducted can lead to later challenges to the reliability, fairness, and objectivity of the identification. Use of the above procedures can minimize such challenges.

D. Recording Identification Results

Show Slide 139

Principle: The record of the outcome of the identification procedure accurately and completely reflects the identification results obtained from the witness.

Policy: When conducting an identification procedure, the investigator should preserve the outcome of the procedure by documenting any identification or nonidentification results obtained from the witness.

Procedure: When conducting an identification procedure, the investigator should—

1. Record both identification and nonidentification results in writing, including the witness's own words regarding how sure he/she is.

Show Slide 140

2. Ensure results are signed and dated by the witness.

Show Slide 141

3. Ensure that no materials indicating previous identification results are visible to the witness.

4. Ensure that the witness does not write on or mark any materials that will be used in other identification procedures.

   👁 In jurisdictions where it is required that a witness sign the back of a selected photo, ensure that the signed photo is not used in a later identification procedure.

Summary: A complete and accurate record of the outcome of the identification procedure can be a critical document in the investigation and any subsequent court proceedings.

Show Slide 142

Further Reading

Each entry below includes a brief synopsis of the publication's focus to assist trainers and students in selecting material for further study.

Connors, E., T. Lundregan, N. Miller, and T. McEwen, *Convicted by Juries, Exonerated by Science: Case Studies in the Use of DNA Evidence to Establish Innocence After Trial.* **Washington, DC: U.S. Department of Justice, National Institute of Justice, 1996, NCJ 161258.**

This NIJ Research Report describes 28 cases in which DNA evidence was used to exonerate persons who had been convicted at trial. The report notes that 24 of these 28 cases involved mistaken identification by the eyewitness(es). The report is also useful for noting other kinds of evidence that may have contributed to the wrongful convictions.

Cutler, B.L., and S.D. Penrod. *Mistaken Identification: The Eyewitness, Psychology, and the Law.* **New York: Cambridge, 1995.**

This book attempts to address the broad range of issues in eyewitness identification, including crossrace identification, "weapon focus," and other topics.

Dunning, D., and L.B. Stern. "Distinguishing Accurate from Inaccurate Identifications via Inquiries about Decision Processes." *Journal of Personality and Social Psychology* **67 (1994): 818–835.**

This article describes experiments that analyze what witnesses say during their identifications (such as, "the face just 'popped out' from the lineup and that is how I made my identification decision"), as well as how these statements differ among witnesses who made accurate versus mistaken identifications.

Fisher, R.P., and M.L. McCauley. "Information Retrieval: Interviewing Witnesses." In *Psychology and Policing,* **ed. N. Brewer and C. Wilson. Hillsdale, NJ: Erlbaum, 1995: 81–99.**

This chapter examines laboratory and field research conducted as part of the cognitive interview (CI) procedure. It summarizes the major principles underlying the CI technique and indicates its strengths and weaknesses. It also describes what conditions are most and least effective for the CI procedure.

Fisher, R.P., and R.E. Geiselman. *Memory Enhancing Techniques for Investigative Interviewing.* **Springfield, IL: Charles Thomas, 1992.**

This book describes in detail how to conduct the cognitive interview procedure to enhance the recall of cooperative eyewitnesses. Examples of correct and incorrect techniques are provided, along with critiques of sample interviews.

Fisher, R.P., R.E. Geiselman, and D.S. Raymond. "Critical Analysis of Police Interview Techniques." *Journal of Police Science and Administration* **15 (1987): 177–185.**

This article describes typical police interview procedures with cooperative witnesses and notes the most common types of errors made by police interviewers. Suggestions are made to improve police interviewing skills.

Geiselman, R.E., and R.P. Fisher. "Ten Years of Cognitive Interviewing." In *Intersections in Basic and Applied Memory Research*, ed. D. Payne and F. Conrad. Mahwah, NJ: Erlbaum, 1997: 291–310.

This chapter summarizes the scientific research used to develop and test the cognitive interview (CI) procedure and also describes instances in which the CI was implemented to solve specific criminal cases.

Lindsay, R.C.L., and G.L. Wells. "Improving Eyewitness Identification From Lineups: Simultaneous Versus Sequential Lineup Presentations." *Journal of Applied Psychology* 70 (1985): 556–564.

This article describes an experiment that compared the simultaneous lineup procedure with the sequential lineup procedure. It explains the research methods used and the psychological principles that make each procedure different.

Loftus, E.F., and J. Doyle. *Eyewitness Testimony: Civil and Criminal*, 3d ed. Charlottesville, VA: Lexis Law Publishing, 1997.

This practice-oriented book, used frequently by defense lawyers, addresses eyewitness reliability and includes references to psychological studies and case law. Issues in expert testimony are discussed extensively.

Malpass, R.S., and R.C.L. Lindsay. "Measuring Lineup Fairness." *Applied Cognitive Psychology* 13 (1999): S1–S8.

This article, the first in a special issue of *Applied Cognitive Psychology* on the topic of lineup fairness, briefly reviews the history of lineup fairness measures. It also provides a simple introduction to quantitative evaluation of lineups and supplies references to other articles that will allow the reader to develop evaluation procedures for his/her own use.

Ross, D.F., J.D. Read, and M.P. Toglia, eds. *Adult Eyewitness Testimony: Current Trends and Developments*. New York: Cambridge, 1994.

This book examines broad issues in eyewitness identification. Different researchers author each of the 18 chapters. Included are chapters on "earwitnesses" (voice identification), distinctions between live and video lineups, the role that personality can play in eyewitness testimony, and how jurors evaluate eyewitness testimony. A chapter by Wells et al., "Recommendations for Conducting Lineups," (pp. 223–244) details various procedures for conducting lineups and the rationale for several general propositions.

Scheck, B., P. Neufeld, and J. Dwyer. *Actual Innocence: Five Days to Execution and Other Dispatches from the Wrongly Convicted*. New York: Doubleday, 2000.

This engaging and helpful book bluntly recognizes the need to improve investigative techniques in the area of eyewitness evidence to ensure that innocent people are not wrongly accused or convicted. It notes that improvement should come from all areas of the criminal justice system—prosecutors, defense attorneys, judges, and law enforcement personnel. Police officials are encouraged to disregard any perceived biases on the part of the authors and digest the book's message.

Sporer, S.L. "Eyewitness Identification Accuracy, Confidence, and Decision Times in Simultaneous and Sequential Lineups." *Journal of Applied Psychology* **78 (1993): 22–33.**

This article provides support for the advantage that sequential lineups have shown in protecting innocent suspects from being falsely identified.

Sporer, S.L., R.S. Malpass, and G. Koehnken, eds. *Psychological Issues in Eyewitness Identification.* **Hillsdale, NJ: Erlbaum, 1996.**

This book contains 12 chapters written by 16 researchers in the field of eyewitness evidence. Topics covered include legal aspects; effects of witness, target, and situational factors; person descriptions; face recall; voice identification; the "other race" effect; enhancing eyewitness memory; forensic applications of eyewitness research; identification evidence from children; elderly witnesses; and the logic and methodology of experimental research in eyewitness psychology. The book was written to appeal to a wide range of professionals in the criminal justice system.

Wells, G.L. "What Do We Know about Eyewitness Identification?" *American Psychologist* **48 (5) (1993): 553–571.**

This article reviews several principles of eyewitness identification that are widely accepted in psychology. It focuses on the concept of relative judgments as a cause of mistaken identification, as well as problems relating to eyewitness confidence.

Wells, G.L., M.R. Leippe, and T.M. Ostrom. "Guidelines for Empirically Assessing the Fairness of a Lineup." *Law and Human Behavior* **3 (1979): 285–293.**

This article describes the most common way that scientific psychologists assess whether the fillers in a lineup are adequate. It expands on such topics as the "mock witness" method and introduces the concept of "functional lineup size."

Wells, G.L., M. Small, S.D. Penrod, R.S. Malpass, S.M. Fulero, and C.A.E. Brimacombe. "Eyewitness Identification Procedures: Recommendations for Lineups and Photospreads." *Law and Human Behavior* **22 (1998): 603–647.**

This is an official paper of the American Psychology-Law Society that represents broad agreement among eyewitness researchers as to how lineups should be conducted. It includes an analysis of the first 40 DNA exoneration cases, which show that 36 of the 40 wrongly convicted people had been convicted primarily because of eyewitness misidentification.

Wells, G.L., S.M. Rydell, and E.P. Seelau. "On the Selection of Distractors for Eyewitness Lineups." *Journal of Applied Psychology* **78 (1993): 835–844.**

This article explains how researchers conduct experiments on eyewitness identification. It discusses in detail the reason why lineup fillers should be selected using the description given by the eyewitness and why they should not be selected merely to look like the suspect. It also presents a major experiment that tests various ways to select fillers.

About the National Institute of Justice

NIJ is the research, development, and evaluation agency of the U.S. Department of Justice. The Institute provides objective, independent, evidence-based knowledge and tools to enhance the administration of justice and public safety. NIJ's principal authorities are derived from the Omnibus Crime Control and Safe Streets Act of 1968, as amended (see 42 U.S.C. §§ 3721–3723).

The NIJ Director is appointed by the President and confirmed by the Senate. The Director establishes the Institute's objectives, guided by the priorities of the Office of Justice Programs, the U.S. Department of Justice, and the needs of the field. The Institute actively solicits the views of criminal justice and other professionals and researchers to inform its search for the knowledge and tools to guide policy and practice.

To find out more about the National Institute of Justice, please contact:

National Criminal Justice
  Reference Service
P.O. Box 6000
Rockville, MD 20849–6000
800–851–3420
e-mail: *askncjrs@ncjrs.org*

Strategic Goals

NIJ has seven strategic goals grouped into three categories:

Creating relevant knowledge and tools

1. Partner with State and local practitioners and policymakers to identify social science research and technology needs.
2. Create scientific, relevant, and reliable knowledge—with a particular emphasis on terrorism, violent crime, drugs and crime, cost-effectiveness, and community-based efforts—to enhance the administration of justice and public safety.
3. Develop affordable and effective tools and technologies to enhance the administration of justice and public safety.

Dissemination

4. Disseminate relevant knowledge and information to practitioners and policymakers in an understandable, timely, and concise manner.
5. Act as an honest broker to identify the information, tools, and technologies that respond to the needs of stakeholders.

Agency management

6. Practice fairness and openness in the research and development process.
7. Ensure professionalism, excellence, accountability, cost-effectiveness, and integrity in the management and conduct of NIJ activities and programs.

Program Areas

In addressing these strategic challenges, the Institute is involved in the following program areas: crime control and prevention, including policing; drugs and crime; justice systems and offender behavior, including corrections; violence and victimization; communications and information technologies; critical incident response; investigative and forensic sciences, including DNA; less-than-lethal technologies; officer protection; education and training technologies; testing and standards; technology assistance to law enforcement and corrections agencies; field testing of promising programs; and international crime control.

In addition to sponsoring research and development and technology assistance, NIJ evaluates programs, policies, and technologies. NIJ communicates its research and evaluation findings through conferences and print and electronic media.

www.ingramcontent.com/pod-product-compliance
Lightning Source LLC
Chambersburg PA
CBHW081223170526
45165CB00009B/2921